Genetic Approaches to Noncommunicable Diseases

Springer
Berlin
Heidelberg
New York
Barcelona
Budapest
Hong Kong
London
Milan
Paris
Santa Clara
Singapore
Tokyo

K. Berg V. Boulyjenkov
Y. Christen (Eds.)

Genetic Approaches to Noncommunicable Diseases

With 19 Figures and 19 Tables

Springer

BERG, KARE, Prof., Dr.
University of Oslo
Institute of Medical Genetics
P.O. Box 1036 Blindern
0315 Oslo
Norway

BOULYJENKOV, VICTOR, Dr.
World Health Organization
Human Genetics Programme
1211 Geneva 27
Switzerland

CHRISTEN, YVES, Dr.
Fondation IPSEN
24, rue Erlanger
75781 Paris
France

Library of Congress Cataloging-in-Publication Data. Genetic approaches to noncommunicable diseases/K. Berg, V. Boulyjenkov, Y. Christen (eds.). p. cm. Includes bibliographical references and index.

ISBN-13: 978-3-642-64662-1 e-ISBN-13: 978-3-642-61028-8
DOI: 10.1007/978-3-642-61028-8

Medical genetics—Congresses. I. Berg, Kare. II. Boulyjenkov, V. (Victor), 1948- . III. Christen, Yves. [DNLM: 1. Hereditary Diseases—genetics—congresses. 2. Hereditary Diseases—prevention & control—congresses. 3. Genetics, Medical—congresses. QZ 50 G32257 1995] RB155.G35926 1995 616'.042-dc20 DNLM/DLC for Library of Congress 95-25145

Softcover reprint of the hardcover 1st edition 1996

Cover design: Springer-Verlag, Design & Production

Typesetting: Thomson Press (India) Ltd., Madras

SPIN: 10482822 27/3136/SPS - 5 4 3 2 1 0 - Printed on acid-free paper

Preface

In recent years, the progress made in the prevention of mortality and morbidity caused by communicable diseases and malnutrition has changed the disease spectrum in both developed and, particularly developing countries. As a result, noncommunicable diseases, including genetic disorders, have achieved considerable importance in public health. Furthermore, it is now evident that inherited predisposition is important in a number of common diseases that occur in later life, such as atherosclerosis, coronary heart disease, hypertension, diabetes mellitus, and in some rheumatic, oncological, and mental illnesses that appear at an early stage and develop into severe handicaps in predisposed people.

Rapid advances in gene mapping concerned with international human genome research make it almost certain that the use of new genetic knowledge will dramatically increase the requirement for genetic approaches in the control of a wide spectrum of diseases, and will provide possibilities for their prevention and treatment in the form of changes in lifestyle, diet modification, periodic check-ups, or the administration of gene therapy. It appears that one of the main problems in delivering genetics services is the difficulty involved in informing the health profession and the community of the real significance of genetic problems. There is, therefore, a need for international collaboration in improving genetic health education at all levels and in improving health through genetic approaches.

In order to achieve this goal, the World Health Organisation (WHO) Human Genetics Programme has been cooperating with the Fondation IPSEN in the last decade. Our collaboration has been developed with regard to improving the diagnosis and prevention of some genetically determined diseases, as well as the education of health care workers. These issues are of great importance to the WHO and the Fondation IPSEN and have been discussed extensively at joint meetings. This particular meeting, held in St. Petersburg (Russia) on 5-6 December 1994, was devoted to actual and potential applications of new genetic technology for health improvement, which continually raise bioethical and scientific issues. The papers presented during the meeting were considered for further revision of the problems, and it is hoped that public awareness of medical genetics will be enhanced.

Although the papers included in this volume have been produced within the framework of WHO/Fondation IPSEN's review of knowledge, they express the views of the individual authors rather than a consensus of the participants at the meeting. The views do not necessarily represent the decision for stated policy of the WHO or the Fondation IPSEN.

The editors wish to express their appreciation to Mrs. Jacqueline Mervaillie for the successful organization of the meeting and to Mrs. Mary-Lynn Gage for her editorial assistance.

K. Berg • V. Boulyjenkov • Y. Christen

Contents

Contributors

Baranov, V.S.
Institute of Obstetrics and Gynecology, Mendeleevskaya line 3, St. Petersburg 199034, Russia

Beisiegel, U.
Medizinische Klinik, Universitätskrankenhaus Eppendorf, Martinistr. 52, 20246 Hamburg, Germany

Berg, K.
Institute of Medical Genetics, University of Oslo and Department of Medical Genetics, Ulleval University Hospital, P.O. Box 1036, Blindern, 0315 Oslo, Norway

Betterridge, D.J.
Department of Medicine, UCLMS, 5th Floor, Jules Thorn Institute, Middlesex Hospital, Mortimer Street, London WIN 8AA, UK

Boulyjenkov, V.
Human Genetics Programme, World Health Organization, 1211 Geneva 27, Switzerland

Cookson, W.
Nuffield Department of Medicine, John Radcliffe Hospital, Oxford OX3 9DU, Great Britain

Corvol, P.
INSERM U 36, Collège de France, 3 rue d'Ulm, 75005 Paris, France

Czeizel, A.E.
Department of Human Genetics and Teratology, National Institute of Hygiene, 1097 Budapest, Hungary

Day, I.N.M.
Centre for Genetics of Cardiovascular Disorders, University College, London Medical School, The Rayne Institute, 5 University Street, London WCIE 6JJ, UK

Dedieu, J.-F.
CNRS URA 1301/Rhone-Poulenc Rorer, Laboratoire des Virus Oncogènes, Institut Gustave Roussy, Rue Camille Desmoulins, 94805 Villejuif Cedex, France

Defesche, J.J.P.
Foundation for the Identification of Persons with Inherited Hypercholesterolemia, Paasheuvelweg 15, 1105 BE Amsterdam, The Netherlands

Denèfle, P.
Rhone-Poulenc Rorer, CRVA, Batiment Monod, 13 Quai Jules Guesde, 94403 Vitry sur Seine Cedex, France

Eriksson, M.
Huddinge University Hospital, Medical Department, M54, 14186 Huddinge, Sweden

Faergeman, O.
Department of Medicine, Aarhus Amtssygehus University Hospital, Tage Hansens Gade 2, 8000 Aarhus C, Denmark

Hamilton-Craig, I.
Little Stirling House, 7 E. Pallant Street, North Adelaide, South Australia 5006, Australia

Hanis, C.L.
Human Genetics Center, The University of Texas Health Science Center at Houston, P.O. Box 20334, Houston, Texas 77225, USA

Hayden, M.R.
Department of Medical Genetics, University of British Columbia, Room 416, 2125 East Mall, Vancouver, British Columbia V6T IZ4, Canada

Hegele, R.A.
St. Michael's Hospital, 30 Bond Street, Toronto, Ontario M5B 1W8, Canada

Humphries, S.E.
Centre for Genetics of Cardiovascular Disorders, University College, London Medical School, The Rayne Institute, 5 University Street, London WCIE 6JJ, UK

Illingworth, R.
Oregon Health Sciences University, 3181 Parnassus, Room L1337 (13th Floor), UCSF School of Medicine, San Francisco, California, USA

Ivanov, V.I.
Research Centre for Medical Genetics, 1 Moskvorechie Street, 115478 Moscow, Russia

Jeck, T.
Department of Endocrinology, Kantonsspital, 4031 Basel, Switzerland

Jeunemaitre, X.
INSERM U 36, Collège de France, 3 rue d'Ulm, 75005 Paris, France

Kane, J.
Cardiovascular Research Institute, Long Hospital, 505 Parnassus, Room L1337 (13th Floor), UCSF School of Medicine, L465, Portland, Oregon, USA

Kastelein, J.C.
STOEH-FIPIH, Paasheuvelweg 15, 1005 BE Amsterdam, The Netherlands

Keller, C.
Medizinische Poliklinik, University of Munich, Pettenkoferstr. 8a, 80336 München, Germany

Keller, U.
Endocrinology, Kantonsspital, 4031 Basel, Switzerland

Klonjkowski, B.
CNRS URA 1301/Rhone-Poulenc Rorer, Laboratoire des Virus Oncogènes, Institut Gustave Roussy, Rue Camille Desmoulins, 94805 Villejuif Cedex, France

Knoppers, B.M.
Université de Montreal, Fac. de Droit, Centre de Recherche en Droit Public, C.P. 6128, Succursale Centre-Ville, Montreal, Quebec H3C 3J7, Canada

Kostner, G.M.
Institut of Medical Biochemistry, University of Graz,
Harrachgasse 21, 8010 Graz, Austria

Kremer, E.J.
CNRS URA 1301/Rhone-Poulenc Rorer, Laboratoire des Virus Oncogènes, Institut Gustave Roussy, Rue Camille Desmoulins, 94805 Villejuif Cedex, France

Kwiterovich, P.O.
John Hopkins University, Lipid Clinic, 550 N. Broadway, Suite 308, Baltimore, Maryland 21205, USA

Lakich, M.M.
CNRS URA 1301/Rhone-Poulenc Rorer, Laboratoire des Virus Oncogènes, Institut Gustave Roussy, Rue Camille Desmoulins, 94805 Villejuif Cedex, France

Latta, M.
CNRS URA 1301/Rhone-Poulenc Rorer, Laboratoire des Virus Oncogènes, Institut Gustave Roussy, Rue Camille Desmoulins, 94805 Villejuif Cedex, France

Lees, R.S.
Boston Heart Foundation, 139 Main Street, Cambridge, Massachusetts 02142, USA

Leitersdorf, E.
Center for Research, Prevention and Treatment of Atherosclerosis, Division of Medicine, Hadassah University Hospital, P.O. Box 12221, 91120 Jerusalem, Israel

Leren, T.P.
Department of Medical Genetics, Ullevål University Hospital, P.O. Box 1036 Blindern, 0315 Oslo, Norway

Mallet, J.
Laboratoire de Neurobiologie Cellulaire et Moléculaire, UMR 9923, CNRS, 91198 Gif-sur-Yvette, France

Marais, A.D.
Lapid Laboratory, H47 Old Groote Schuur Hospital, Internal Medicine, University of Cape Town Medical School, Observatory 7925, Cape Town, South Africa

Meloni, R.
Laboratoire de Neurobiologie Cellulaire et Moléculaire,
UMR 9923, CNRS, 91198 Gif-sur-Yvette, France

Miserez, A.R.
Department of Molecular Genetics, University of Texas,
South-western Medical Center, 5323 Harry Hines Boulevard,
Dallas, Texas 75235, USA

Mulvihill, J.J.
University of Pittsburgh, Department of Human Genetics,
A300 Grabtree Hall, 130 DeSoto Street, Pittsburgh,
Philadelphia 14261, USA

Orsini, C.
CNRS URA 1301/Rhone-Poulenc Rorer, Laboratoire des Virus
Oncogènes, Institut Gustave Roussy, Rue Camille Desmoulins,
94805 Villejuif Cedex, France

Ose, L.
Lipidklinikken, Department of Medicine, Rikshospitalet,
Pilestredet 32, 0027 Oslo, Norway

Perricaudet, M.
CNRS URA 1301/Rhone-Poulenc Rorer, Laboratoire des Virus
Oncogènes, Institut Gustave Roussy, Rue Camille Desmoulins,
94805 Villejuif Cedex, France

Pimstone, S.N.
Department of Medical Genetics, University of British Columbia,
Room 416, 2125 East Mall, Vancouver, British Columbia V6T 1Z4,
Canada

Prost, E.
CNRS URA 1301/Rhone-Poulenc Rorer, Laboratoire des Virus
Oncogènes, Institut Gustave Roussy, Rue Camille Desmoulins,
94805 Villejuif Cedex, France

Raal, F.J.
Department of Medicine, Medical School,
University of Witwatersrand, 7 York Road, Parktown 2193,
Johannesburg, South Africa

Schuster, H.
Franz-Volhard-Klinik, Max-Delbrück-Centrum, Wiltbergstr. 50,
13122 Berlin, Germany

Seftel, H.C.
Department of Medicine, Medical School,
University of Witwatersrand, 7 York Road, Parktown 2193,
Johannesburg, South Africa

Soubrier, F.
INSERM U 36, Collège de France, 3 rue d'Ulm, 75005 Paris, France

Stein, E.
Cardiovascular Research Center, 2350 Auburn Avenue, Cincinnati,
Ohio 45219, USA

Steinhagen-Thiessen, E.
Universitätsklinikum Rudolf Virchow, Sophie-Charlotten-Str. 115,
14059 Berlin, Germany

Verma, I.C.
WHO Collaborating Centre in Genetics, Genetic Unit,
Department of Pediatrics, All India Institute of Medical Sciences,
New Delhi 110029, India

Vigne, E.
CNRS URA 1301/Rhone-Poulenc Rorer, Laboratoire des Virus
Oncogènes, Institut Gustave Roussy, Rue Camille Desmoulins,
94805 Villejuif Cedex, France

Williams, R.R.
Cardiovascular Genetis, University of Utah, 410 Chipeta Way,
Room 161, Salt Lake City, Utah 84108, USA

Yeh, P.
CNRS URA 1301/Rhone-Poulenc Rorer, Laboratoire des Virus
Oncogènes, Institut Gustave Roussy, Rue Camille Desmoulins,
94805 Villejuif Cedex, France

Genetics and Environment in Common Diseases

V.I. IVANOV

A Bit of History

More than a century ago, Sir Francis Galton (1889) posed a query, "nature or nurture?," aimed at estimating the "relative powers" of these two factors in human development. For several decades Galton's query was of paramount importance in directing human genetic research on an exact quantitative course, especially when common morbid traits were considered.

In both the origin and manifestation of the vast majority of common disorders, hereditary and environmental factors each play their respective obligatory roles. In other words, the history of a common human disease is usually the history of both the genetic predisposition *and*, not *or*, provocative environmental factors. The origin of this notion can be easily traced back to the 1920s and 1930s, when animal and plant phenogeneticists of the time formulated their first theories about genotype-phenotype interrelations in organism development.

In my judgement, which may be formed by my professional origin, one of the earliest and most comprehensive models of phenotypic manifestation of a genotype was elaborated by S.S. Tschetverikoff (also spelled Chetverikov) and two of his students, N.W. Timofeeff-Ressovsky and B.L. Astauroff. According to their personal communications, the discussions on genotype-phenotype relations took place at Tschetverikoffs' tea parties as early as 1922–1924. In Russian these discussions were nicknamed "дрозсоор" a shortening of "Совместное орание дрозофильщиков", which translates as "Drosophilists' joint shouting" in English.

In one of his earliest publications of 1925, Timofeeff-Ressovsky reported on quantitative and qualitative variation in gene expression in *Drosophila*. In his 1926 work, co-authored with O. Vogt, the terms "penetrance," "expressivity," and "specificity"-designating respective measures of variation in gene expression-were first introduced. The authors also discussed the application of the experimental phenogenetic data and concepts to human disease classification.

Even in early studies, gene penetrance and expressivity were found to be influenced by both the genotypic background and the ambient conditions. Therefore, Tschetverikoff (1926) suggested the need to distinguish between the

K. Berg, V. Boulyjenkov, Y. Christen (Eds.)
Genetic Approaches to Noncommunicable Diseases

outer environment of an organism (*medium externum*) and the genotypic environment (genetic background, milieu genetique, *medium genotypicum*) of the gene in question. Somewhat later, by examining the bilateral variation in halters-to-wing homoeotic transformation in tetraptera mutants of *Drosophila*, Astauroff inferred the existence of one more medium, namely the intraorganismal variation in development that he designated as internal medium (*medium internum*). Summarizing his own and other authors' results, Astauroff (1930) produced a general formula of organism development:

$$a = F(p, g, e, i)$$

where a is a resulting trait, p = plasmatype, g = genotype, e = external medium, and i = internal medium.

A similar, graphic model was gradually developed by Timofeeff-Ressovsky, beginning in the above-mentioned paper with Vogt (1926), up to a completed version published in 1940 and later (cf. Timofeeff-Ressovsky and Ivanov 1966), which is shown in Fig. 1.

Neither of these models specified whether the effects of the factors involved were additively totalled up or if a sort of interaction existed between them. In the 1970s we carried out an extensive study in *Drosophila* on the mode of action of some dozen homoeotic and other genes being combined in pairs and triplet combined in the same stocks (e.g. Boulyjenkov and Ivanov 1977, 1978; Kaurov et al. 1978; Boulyjenkov 1979; Ivanov 1979; Khusnutdinova et al. 1981, 1982). The total data obtained and shown in Table 1 brought us to the unequivocal conclusion that the joint manifestation of the genes studied was interactive.

In general, the accumulated facts and elaborated models could already describe how, in the wonderful and very complicated ontogeny of *Metabiota*, specific events take place at a specific time and at the specific location (Timofeeff-Ressovsky and Ivanov 1966; Timofeeff-Ressovsky et al. 1977; Ivanov 1991).

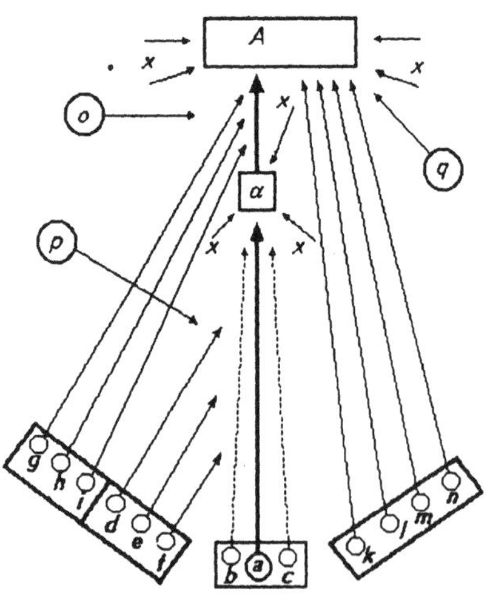

Fig. 1. Formal scheme of gene expression (from Timofeeff-Ressovsky 1940). The manifestation of a gene (a) in definite trait (A) through intermediate stage (α) is influenced by other genes of the genome (b–n) and intraorganismal (p–q) and external (x) factors

Thus, the basic features of the phenomenology of gene manifestation were more or less outlined. Further studies have shown that general inferences from animal and plant developmental genetics can be extrapolated to the formation of both normal and morbid human hereditary traits (e.g.,Bochkov and Ivanov 1981, 1991; Ivanov 1982, 1992; Lazyuk et al. 1982; Bochkov et al. 1984, 1988).

Non-Communicable Diseases Today

This section is substantially based on the World Health Organization document, "Research for Health" (1993).

An investigation recently carried out by the Advisory Committee on Health Research and aimed to "develop further a clearly enunciated health research strategy for WHO in order to translate the research goals, priorities and programmes into coherent and coordinated action in support of health for all" has shown that non-communicable disorders become major causes of disability and death in most countries of the world.

The non-communicable disorders include prenatal diseases determined at fertilization (e.g., single gene defects and chromosomal aberrations), prenatal diseases determined after fertilization (e.g.,the result of hazards associated with early embryonic development), postnatal diseases due to deficiencies and hazards (diseases of poverty), and postnatal diseases due to maladaptation ("diseases of affluence").

Disease is not an inescapable attribute of the human condition; except when determined at or soon after fertilization, it results essentially from unhealthy ways of living and can be prevented if those ways can be changed.

The frontiers of knowledge are constantly being pushed back as a result of advances in research relevant to health care and health services development, particularly in cell biology, including genetics, biotechnology, the neurosciences

Table 1. Interaction of homoeotic genes in organ transformation in Drosophila. (From Ivanov 1979)

Genotype	Transformed organ			
	Proboscis	Antenna		Second and Third legs
		Proximal	Distal	
pb/pb	1	0	0	0
Antp/ + Ns/Ns	0	1	0.42	0
ss^a/ss^a	0	0	1	0
Pc/ +	0	0	0	1
pb Ns/pb Ns	2.20	1.67	1.67	0
pb ss^a/pb +	2.31	0	1.03	-
pb Pc/pb +	-	-	-	-
Antp ss^a/ + ss^a	0	2.13	1.84	0
Antp/Pc	0	4.73	1.97	0.64
Pc $ss^{a/}$ + ss^a	-	0	1.16	2.38

and the physical sciences. Other potential contributions of new knowledge to improve the health of individuals and communities may come from the biological, agricultural, physical, social and environmental sciences.

Therefore, research strategy aimed at the efficient control of non-communicable diseases requires investigation into the genetic, environmental, and behavioural influences that have led to these diseases that are now predominant in developed countries and beginning to rise in the developing world. In some, the major exogenous influences (tobacco, alcohol, occupational hazards, pollution, etc.) are already known, and the research required is predominantly concerned with heredity and behaviour; in others, the influences are unknown, and research is needed into disease origins.

Genetic Heterogeneity of Diseases

Many genetic disorders are caused by mutations at two or more loci. Conversely, similarities in the clinical presentation of two apparently distinct disorders may suggest that the different phenotypes represent allelic heterogeneity at a single locus (Suthers and Davies 1992).

Genetic linkage studies in families with Duchenne and Becker muscular dystrophies lent support to the clinicians' notion that the two disorders are allelic (Kingston et al. 1984). Subsequent identification of mutations in the dystrophin gene have confirmed this view (Koenig et al. 1987; Hoffman et al. 1988).

In autosomal recessive proximal spinal muscular atrophy (SMA), the anterior horn cells of the spinal cord are affected with consequent muscle weakness, but nothing is known of the pathogenesis (Suthers and Davies 1992). The patients usually present with one of the three forms of the disease. In the most severe type 1 (Werdnig – Hoffmann disease), the onset of symptoms is in the first six months of life; the child never sits, and death usually occurs by the age of 2 years. The intermediate type II has a later onset, but the affected child never walks unaided. In the comparatively mild type III (Kugelberg – Wellander disease), affected individuals maintain independent ambulation. The three types of SMA have been mapped to chromosome 5 (e.g., Brzustowicz et al. 1990). In carefully evaluated multiplex pedigrees there is no evidence of either incomplete penetrance or locus heterogeneity (Dubowitz 1991). Although the clinical features of SMA are usually consistent among the affected individuals in a pedigree, there are several reported pedigrees in which wide variation in the severity of muscle weakness was observed. In such cases other genetic or environmental factors must be responsible for the variability in phenotype.

In a vast Azerbeidjanian pedigree of 399 members that could be traced to a single founder couple through four consecutive generations and included 92 persons affected with Ehlers-Danlos syndrome (Prytkov et al. 1984; Blinnikova 1985), three types of the syndrome were actually observed and, in some cases, the type of the disease in the offspring was not the same as in his/her parents and ancestors (Fig. 2, Table 2).

Thus, heterogeneity of genetic diseases is in itself heterogeneous and may be caused by diverse genetic events, variable environmental influences, and the complicated interactions of the two.

The same holds true, to an even greater extent, when common non-communicable diseases are considered. In a recent genome-wide search for human insulin-dependent diabetes mellitus susceptibility genes in 96 families (Davies et al. 1994), in addition to *IDDM1* gene (in the major histocompatibility complex on chromosome 6p21) and *IDDM2* (in the insulin gene region on chromosome 11p15), linkages to chromosome 11q (*IDDM4*) and 6q (*IDDM5*) were confirmed. The authors suggest a polygenic inheritance of the diabetes type 1 with a major locus at the major histocompatibility complex.

Genes and Environment

In this section the genome-environment interactions are exemplified by data on carcinogenesis mostly compiled from a review by Shields and Harris (1991).

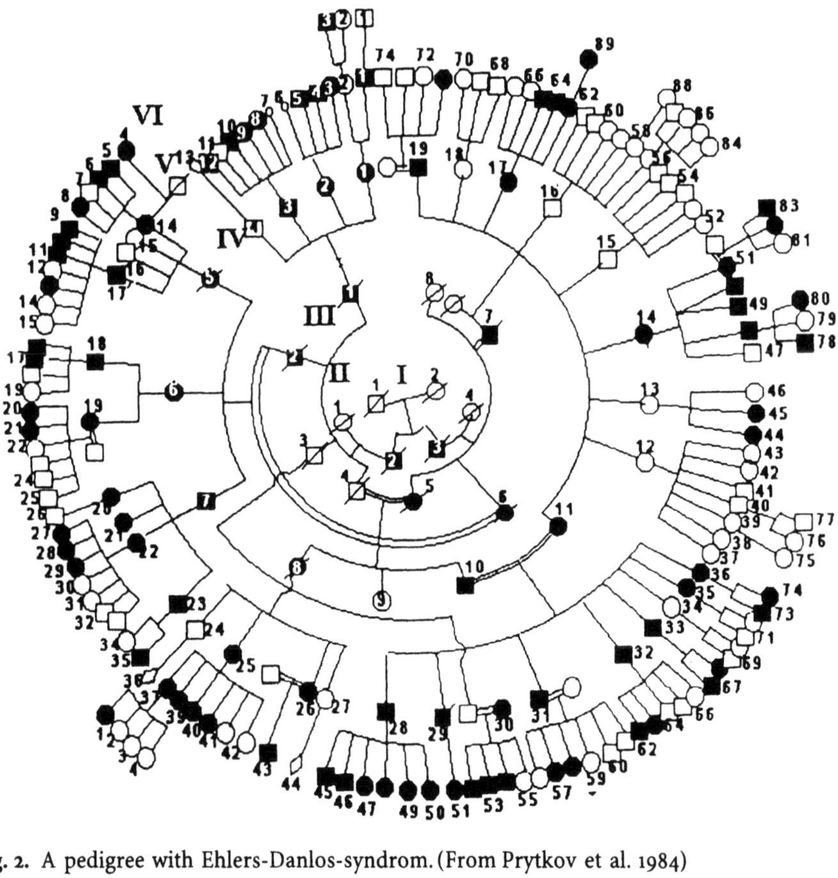

Fig. 2. A pedigree with Ehlers-Danlos-syndrom. (From Prytkov et al. 1984)

Table 2. Types of Ehlers-Danlos syndrome in parents and children from a pedigree. (From Blinnikova 1985)

Children	Parents		
	I	II	III
I	30	7	-
II	12	9	-
III	8	3	-

Carcinogenesis is a multistage process of normal growth, differentiation, and development gone awry. It is driven by spontaneous and carcinogen-induced genetic and epigenetic events. Figure 3 presents a simplified scheme.

Tumor initiation involves the direct effects of carcinogenic agents on DNA, mutations, and altered gene expression. Tumor promotion further involves an "initiated" cellular clone that may also be affected by growth factors that control signal transduction. During this process, progressive phenotypic changes and genomic instability occur (aneuploidy, mutations, or gene amplification). These genetic changes enhance the probability of initiated cells transforming into a malignant neoplasm. Ultimately, tumor cells can disseminate through vessels, invading distant tissues and establishing metastatic colonies. In these stages specific roles are played by proto-oncogenes, tumor suppressor genes, mutations and other damage to DNA, repair of the latter, etc.

In almost every step of the multistage process of carcinogenesis, person-to-person differences in cancer susceptibility can be found. Interindividual differences for particular traits can be acquired or inherited. Inheritance of the ability to metabolise debrisoquin sulfate, an antihypertensive medication, is correlated with the cancer risk. Another example is the aryl hydrocarbon hydroxylase enzyme involved in the metabolism of polycyclic aromatic hydrocarbons. Activity varies in lung tissue but can also be inducible on exposure to agents such as tobacco smoke. Induction is notably higher in patients with lung cancer than in non-cancer controls.

The multiple stages of carcinogenesis are best exemplified by a model of human colorectal tumorigenesis (Fearon and Vogelstein 1990). In the early stages, and apparently more commonly in patients with familial polyposis, the loss or inactivation of a candidate tumor suppressor gene, MCC located on chromosome 5q, is associated with cellular hyperproliferation. More advanced tumors involve oncogens and tumor suppressor genes not observed in early adenomas. These include Ki-ras mutations on chromosome 12, mutation of p53 tumor suppressor genes on chromosome 17, and a deletion of DCC, the putative tumor suppressor gene on chromosome 18q that may be involved in cell-to-cell adhesion and possibly metastasis. Thus, it appears that at least six genetic events occur in the development of colorectal carcinoma.

Genetic damage from carcinogens depends on exposure, absorption, metabolism, DNA repair, etc., each of which can be affected by host factors.

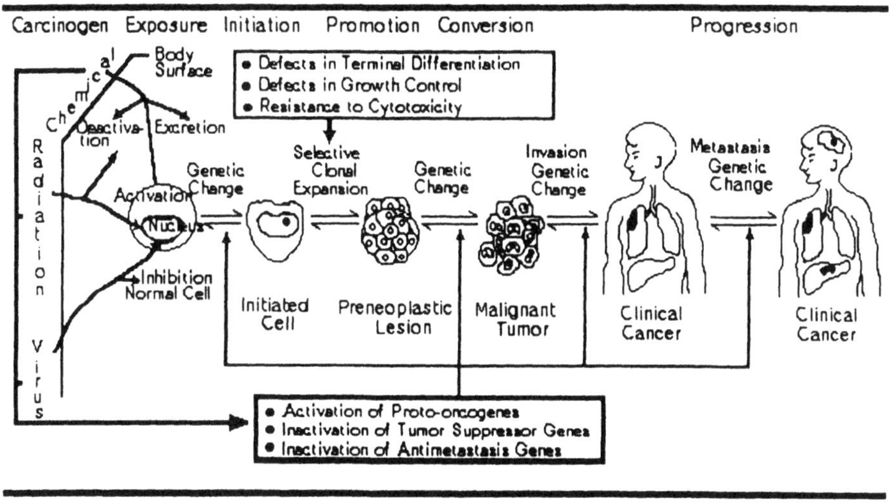

Fig. 3. Multistage process of carcinogenesis. (From Shields and Harris 1991)

Exposures to polycyclic aromatic hydrocarbon (PAH) compounds are associated with an increased risk of lung and skin cancer. Industrial pollution, fossil fuels, and tobacco smoke account for the major environmental sources. Dietary exposures also occur, due to overcooked or charcoal-broiled meats. Carcinogenic PAHs are metabolised by cytochrome P-450 monooxygenases, and reactive epoxide intermediates readily form adducts in DNA (Dipple 1994). Adduct levels have been correlated with exposure in coke-oven workers, tobacco consumption, and urban versus rural residences, but decreases during vacation from occupational sources have been observed.

Nitrosoamines and other nitrosocompounds are potential human carcinogens. These compounds readily alkylate DNA and form adducts. Exposure can occur through endogenous formation of nitrosoamines or directly from dietary sources, cosmetics, drugs, tobacco smoke, and household commodities.

Among the best studies potential dietary carcinogens are aflatoxins produced by *Aspergillus flavus* and *Aspergillus parasiticus*. These molds are contaminants of corn, peanuts, sorghum, and other agricultural products. Aflatoxin B_1 consumption has been linked to a mutation in the p53 tumor suppressor gene in hepatocellular carcinomas.

Interactive effects of environmental agents and host factors have been observed (Haugen and Harris 1990), and some are presented in Table 3.

Inheritance plays an important role in host susceptibility to cancer. Family cancer syndromes can lead to up to a 1000-fold increased risk of cancer. Host susceptibility can be manifested through inheritable interindividual differences in metabolism, DNA repair, genomic instability, or altered proto-oncogene or suppressor gene expression.

The study of the cytochrome P-450 CYP2D6 enzyme is among the best examples of inheritable interindividual differences in metabolism. The extensive

Table 3. Interactive Effects of Carcinogen Exposure and Host Factors. (From Shields and Harris 1991)

Type	Example[a]	Associated Tumor
Chemical-host	PAH and CYP2D6	Lung
	Tobacco smoke and CYP2D6	Lung
	Chemotherapy	Leukemia
Physical-host	Asbestos and CYP2D6	Lung
	Sunlight and xeroderma pigmentosum	Skin
	Radiation and RB-deficient genotype	Osteosarcoma
Viral-host	EBV and X-linked immunodeficiency syndrome	Lymphoma

[a]EBV-Epstein-Barr virus; CYP2D6-cytochrome P-450 CYP2D6 metabolic phenotype determined by debrisoquin sulfate administration and measurement of urinary metabolites; PAH-polycyclic aromatic hydrocarbon; RB-retinoblasoma.

metabolic phenotype has been shown to have an interactive effect with occupational exposures to asbestos and PAHs.

An investigation involving the metabolism of isoniazid, commonly used in the treatment of tuberculosis, led to the identification of acetylation phenotypes for the enzyme acetyltransferase. This enzyme, besides metabolising caffeine, also metabolises certain carcinogenic aromatic amines. The slow acetylation phenotype has been associated with increased risk of bladder cancer. In contrast, rapid acetylation is correlated with the risk of colon cancer. A correlation of aromatic amine adducts and acetylator phenotype in tobacco smokers highlights the relationship of acetylation and DNA damage.

Conclusion

Even the few selected examples discussed above leave no reasonable grounds to doubt that, in non-communicable diseases as a whole as well as in hereditary diseases, the resulting clinical manifestations are always due to an interaction of genotypic, organismal and environmental factors. The major point is not to prove this self-evident notion, but to elucidate the actual substances and processes that determine the onset, progression, variation, and curability of a given disease in a given patient.

Acknowledgements. The author is very much obliged to the WHO Advisory Committee on Health Research and to Drs. O.E. Blinnikova, C.C. Harris, A.N. Prytkov et al., and P.G. Shields for extended use of their materials from publications included in the reference list.

References

Astauroff BL (1930) Analyse der erblichen Stoerungsfaelle der bilateralen Symmetrie im Zusamenhang mit der selbstaendigen Variabilitaet aenlicher Strukturen. Ztschr ind Abst. -Vererb 55: 183-262

Blinnikova OE (1985) Characterization and genetic analysis of clinical polymorphism in Ehlers - Danlos syndrome (Russ). Thesis, Institute of Medical Genetics, Moscow

Bochkov NP, Ivanov VI (1981) Genetic mechanisms of organism homeostasis (Russ). In: Gorizontov PD (ed) Homeostasis. 2nd Edn. Meditsina, Moscow, pp 241-255

Bochkov NP, Ivanov VI (1991) Genetic factors of disease chronization (Russ). Clin Med (Russ) 69-3: 15-18

Bochkov NP, Zakharov AF, Ivanov VI (1984) Medical genetics (Russ). Meditsina, Moscow

Bochkov NP, Zakharov AF, Ivanov VI (1988) Medizinische Genetik. Fischer, Jena

Boulyjenkov VE (1979) Genes interaction in *Drosophila* development (Russ). In: Ivanov VI (ed) Animal developmental genetics. The USSR Institute of Scientific and Technical Information, Moscow, pp 154-187 (General Genetics, vol 6)

Boulyjenkov VE, Ivanov VI (1977) Interaction of homoeotic genes *Antennapedia*, *aristapedia*, and *Polycomb* in *Drosophila melanogaster* (Russ). Genetika 13: 1586-1595

Boulyjenkov VE, Ivanov VI (1978) Interaction of Homoeotic Genes *Antennapedia*, *aristapedia*, and *Polycomb* in *Drosophila melanogaster*. Biol Zbl 97: 527-533

Brzustowicz LM, Lehner T, Castilla LH, Penchaszadeh GK, Wilhelmsen KC, Daniels R, Davies KE, Leppert M, Ziter F, Wood D, Dubowitz R, Zer K, Hansunanowa-Petrusewigcz I, Oft J, Munsat TL, Gilliam TC (1990) Genetic mapping of chronic childhood-onset spinal muscular atrophy to chromosome 5q11. 2-13. 3. Nature 344: 540-541

Davies JL, Kawaguchi Y, Bennett ST, Copeman JB, Cordett HJ, Pritchard LE, Reed PW, Gough SCL, Jenkies SC, Palmer SM, Balfour KM, Rowe BR, Farrall M, Barnett AH, Bain SC, Todd JA (1994) A genome-wide search for human type 1 diabetes susceptibility genes. Nature 371: 130-135

Dipple A (1994) Reactions of polycyclic aromatic hydrocarbons with DNA. In: Hemminki K, Dipple A et al. (eds) DNA adducts: identification and biological significance. IARC Sci Publ No 125, Lyon, p 107

Dubowitz V (1991) Chaos in the classification of the spinal muscular atrophies of childhood. Neuromuscular Dis 1: 77-80

Fearon ER, Vogelstein B (1990) A genetic model for colorectal tumorigenesis. Cell 61: 759-767

Galton F (1889) The history of twins as a criterium of the relative powers of nature and nurture. McMillan, London

Haugen A, Harris CC (1990) Interactive effects between viruses and chemical carcinogens. In: Cooper CS, Grover PL (eds) Carcinogenesis and mutagenesis. Handbook of Experimental Pharmacology. Springer, New York, pp 249-268

Hoffman EP, Fishbeck KH, Brown RH, Johnson M, Medori R, Loike JD, Harris JB Waterston R, Brooke M, Specht L, Kupsky W, Chamberlain J, Caskey T, Shapiro F, Kunkel LM (1988) Characterization of dystrophin in muscle-biopsy specimens from patients with Duchenne's or Becker's muscular dystrophy. N Eng J Med 318: 1363-1368

Ivanov VI (1979) Interaction of genes controlling the processes of cells' determination (Russ). In: Zakharov AF (ed) Theoretical problems in medical genetics. The USSR Academy of Medical Sciences, Moscow, pp 100-114

Ivanov VI (1982) Multiple inborn malformations: approaches to elaboration of human developmental genetics (Russ). In: Proc 4th Congress NI Vavilov Society of Geneticists and Breeders of the USSR. Nauka, Moscow, pp 237-238

Ivanov VI (1991) Some ethical aspects of genetic approaches to human health care: a developmental geneticist's point of view. In: Sram RJ, Bulyzhenkov V, Prilipko L, Christen (eds) Ethical issues of molecular genetics in psychiatry. Springer, Berlin Heidelberg New York, pp 57–60
Ivanov VI (1992) Current trends in hereditary diseases: research and management. World Health Organization, Geneva
Ivanov VI (1993) Interaction of genetic and environmental factors in the control of development (Russ). Ontogenez 24: 85–95
Kaurov BA, Ivanov VI, Mglinets VA (1978) Interaction of homoeotic *proboscipedia* gene with dachs and *four jointed* "leg" genes in *Drosophila melanogaster* at different temperatures (Russ). Genetika 14: 306–312
Khusnutdinova EK, Boulyjenkov VE, Ivanov VI (1981, 1982) Genetic control of determination of thoracic and antennal disks in *Drosophila melanogaster* (Russ). Comm 1–2 Genetika 17: 476–491; Comm 3 ibid 17: 2152–2159; Comm 4 ibid 18: 248–254
Kingston HM, Sarfarazi M, Thomas NST, Harper PS (1984) Localization of the Becker muscular dystrophy gene on the short arm of the X chromosome by linkage to cloned DNA sequences. Human Genet 67: 6–17
Koenig M, Hoffman EP, Bertelson CJ, Monaco AP, Feener C, Kunkel LM (1987) Complete cloning of the Duchenne muscular dystrophy (DMD) cDNA and preliminary genomic organization of the DMD gene in normal and affected individuals. Cell 50: 509–517
Lazyuk GI, Ivanov VI, Tolarova M, Czeizel A (1982) Genetics of inborn malformations (Russ). In: Bochkov NP (ed) Perspectives in medical genetics. Meditsina, Moscow, pp 187–240
Li FP (1990) Familial cancer syndromes and clusters. Curr Probl Cancer 14: 73–114
Prytkov AN, Kozlova SI, Sultanova FA, Blinnikova OE, Garkavtsev IV (1984) Genetic analysis of Ehlers-Danlos syndrome in a large pedigree (Russ). Genetika 20: 868–873
Research for Health. Principles, perspectives and strategies (1993). World Health Organization, Geneva
Shields PG, Harris CC (1991) Molecular epidemiology and the genetics of Environmental Cancer. JAMA 266: 681–687
Suthers GK, Davies KE (1992) Phenotypic heterogeneity and the single gene. Am J Hum Genet 50: 887–891
Timofeeff-Ressovsky NW (1925) On phenotypic manifestation of genotype (Russ). J Exp Biol A1: 93–142
Timofeeff-Ressovsky NW (1940) Allgemeine Erscheinungen der Gen-Manifestierung. In: Just G (ed) Handbuch der Erbbiologie des Menschen, Bd 1. Julius. Springer, Berlin, pp 32–72
Timofeeff-Ressovsky NW, Ivanov VI (1966) Some problems in phenogenetics (Russ). In: Alikhanyan SI (ed) Current problems in genetics. Moscow Univ Press, Moscow, pp 114–130
Timofeeff-Ressovsky NW, Vogt O (1926) Ueber idiosomatische Variationsgruppen und ihre Bedeutung fuer die Klassifikation der Krankheiten. Die Naturwiss 108: 1188–1190
Timofeeff-Ressovsky NW, Ginter EK, Ivanov VI (1977) On some problems and goals in phenogenetics (Russ). In: Belyaev DK (ed) Problems in experimental biology. Nauka, Moscow, pp 186–195
Tschetverikoff (Chetverikov) SS (1926) On certain aspects of evolutionary process from the standpoint of modern genetics (Russ). J Exp Biol A2: 3–56
World Health Organization (1993) Research for health: principles, perspectives and strategies. World Health Organization, Geneva

Predictive Testing for Huntington Disease: Lessons for Other Adult Onset Disorders

M.R. HAYDEN

Introduction

Predictive testing for Huntington disease (HD) has now been offered for approximately eight years in Canada (Fox et al. 1989, Babul et al. 1993). Prior to the development of protocols for predictive testing, there were significant concerns about the impact of providing modification of risk for a disorder for which there was no treatment. It was feared that disclosure of results could have significantly adverse effects on the quality of life of participants (Perry 1981).

Over the last eight years, approximately 500 persons have received predictive testing results. We have previously reported the one-year assessment of the psychological functioning of persons who received a modification of risk, using measures psychological distress, including the general severity index of the symptom checklist 90, the Beck depression inventory and the general well-being scale (Wiggins et al. 1992). These were administered before testing and again at intervals of seven to ten days and six, 12, 18 and 24 months after participants received their test results. Compared to baseline, seven to ten days after disclosure those receiving a decreased risk result had a greater sense of well-being ($p < 0.001$), less depression ($p = 0.003$) and distress ($p < 0.001$). In addition, at each subsequent follow-up, including six, 12, 18 and 24 months, the levels of psychiatric distress remained significantly lower than baseline for this group. These results suggest that, for people who receive a decreased risk result, there is generally an improvement of psychological health as evaluated by these measures of psychological status.

A similar trend has been seen for those individuals who have received an increased risk result. In particular, an overall improvement in their scores during the follow-up period was detected on both the general severity index ($p = 0.03$) and the Beck depression inventory ($p < 0.001$). This group also demonstrated significant linear declines on the general severity index ($p = 0.02$) and the Beck depression inventory ($p = 0.18$) over the 12-month and 24-month periods. Therefore, for this particular group receiving an increased risk result has resulted in some improvement in psychological health as evaluated by these measures (Wiggins et al. 1992). However, as we have previously described, within

K. Berg, V. Boulyjenkov, Y. Christen (Eds.)
Genetic Approaches to Noncommunicable Diseases

both the decreased and increased risk groups there was a small proportion of persons who had adverse outcomes (Lam et al. 1988; Bloch et al. 1992; Huggins et al. 1992).

From Research to Service: Opinions of Health Care Providers and Patients

Our research protocol combined educational materials (pamphlets and videotapes), counselling and a number of questionnaires. In addition, patients were encouraged to bring in a personal support system, such as a spouse or close friend, and all patients were given a 24-hour telephone number to call if they required urgent support. As part of a survey of 256 participants in the study, over 95% indicated that they were satisfied with their experience and more than one-third indicated that they were very satisfied. Only 10 individuals felt they were on balance dissatisfied and, of these, five received an uninformative result and three had not wanted testing from the outset. Of the two individuals who received their test result and reported dissatisfaction, both received an increased risk result of developing HD (Copley et al. 1995).

Approximately one-third of the participants felt that their quality of life had improved following predictive testing. In contrast, 25 individuals (2%) felt their quality of life had decreased because of the participation. Interestingly, the effect of participation and quality of life in the present study was highly associated with group membership, with proportionally more individuals in the decreased risk group commenting that their quality of life had improved (48.1%). There were significantly fewer persons in the increased risk group than in the decreased risk group ($p = 0.01$) who felt the quality of life had improved. However, only three people in the increased risk group (4.5%) and one person in the decreased risk group (0.9%) felt their quality of life had deteriorated.

Approximately 83% of patients felt that the 24-hour telephone contact number was an essential part of the service protocol. In contrast, only 7% of the professionals involved viewed this component as essential. This finding is probably due to the perception that the availability of a 24-hour contact number would be adding potential demands to an already overstretched clinical service. However, it is interesting to note that the 24-hour contact number was actually used by very few individuals. The patient's recoginition that support is always available can be seen as a highly economical and effective way to decrease patient anxiety and improve satisfaction with the programme.

Another area of significant difference between participants and professionals providing the service was the view of the role of the general practitioner (GP). Prior studies have shown that GPs are usually willing to participate in predictive testing protocols (Mennie et al. 1990; Thomassen et al. 1993; Thies et al. 1993). However, 21% of patients indicated their GP should not be routinely involved in the predictive testing process. This suggested that, for some patients, predictive testing is viewed as a highly personal matter that they would prefer to keep separate from their doctor and out of their doctor's patient records. Conversely, only 3% of clinicians felt the GP should not be routinely involved. Clinicians may have felt more favourably about involving the GP since they may see it as a way of reducing the counselling load and returning the patient to the professional who is likely to have an understanding of the patient's environment and would be involved in ongoing patient care. However, it is clearly important that clinicians offering predictive testing recognise that some patients have not told, and may not wish to tell, their GP about their involvement. This finding underscores the need to obtain consent from the patient before releasing information to the GP.

Moving from a research to a service protocol, both the providers and consumers of predictive testing may have significant differences and clearly these need to be taken into account when providing predictive testing in the general community.

Laboratory Issues

After the linked marker for HD was discovered in 1983 (Gusella et al. 1993), predictive testing for HD became available using markers closely linked to the HD gene. However, it was recognised at that point that there was a potential error in this assessment, as one was inferring the presence of the mutation by virtue of the phase of a marker that was segregating with the disease in that family. This raised the possibility of error for that assignment, in view of the fact that recombination between the marker and the disease gene might have occurred. The likelihood of recombination is in part dependent upon the distance between the marker and the site for the mutation. For markers closely linked to the HD gene, a figure of 2% error rate was usually used in the calculation.

After the cloning of the HD gene (HD Collaborative Group 1993) and the recognition that HD was due to expansion of a CAG trinucleotide within a novel gene in almost all patients (Kremer et al. 1994), it was possible to reassess individuals in an effort to see whether the initial designation of risk was correct. In addition, at any phase of the study, other errors in assignment of risk could occur due to human error, including mixing up the samples after withdrawal of the blood, or laboratory error.

A total of approximately 530 predictive testing results have been provided in Canada. Of these, we have now recognised that six test results were in error. Reasons for these errors were: laboratory or clinical error (two), recombination (one), changes in determination of which phase of the marker segregated with the disease due to new information in the family (two) and research developments particularly related to new information about the origin of new mutations (one). In an effort to prevent human error (two instances), new guidelines for quality control within the laboratory have been established. These guidelines are applicable to many other illnesses and to DNA banking procedures in general.

Quality Control Procedures for DNA Banking for HD

Paper work

Upon receiving a blood sample, the DNA Bank Coordinator first checks that the name on the blood tube corresponds to the name on the requisition form and on the DNA banking and consent forms. This check consists of double-checking the spelling of both the first and last names and the accuracy of the date of birth (e.g., confirming that 5-4-32 is the 5th of April, 1932, and not the 4th of May, 1932). Sometimes this check requires calls to genetic centres in Canada and across the world. Any discrepancy in the date of birth or spelling of names is resolved before the banking process begins.

In addition, a pedigree check is performed. The name, risk status, and date of birth on the pedigree must correspond with the information found on the blood tubes and on the DNA banking and consent forms.

Once the subject has been linked to a pedigree (or been given a new pedigree number if it is not possible to connect him or her to an existing pedigree), all relevant information about the person is recorded in the DNA banking receipt book, including: the subject's full name, sex, sample code, and pedigree number; the date the sample was received; the subject's risk status; how much blood was received; and the name of a contact. If the subject requesting DNA banking or predictive testing has so requested on his or her forms, the contact person is written a thank-you note stating that the blood has arrived in good condition.

Each subject is assigned a unique sample code, which consists of the month and year the sample was taken in, plus a numerical tag that reflects how many samples have already been taken in that month. For instance, the code 94-07-023 would indicate that the sample is from the 23rd person to have DNA banked in July of 1994. If another sample arrived in that month, it would be coded 94-07-024. The procedure ensures that there will be no duplicate sample codes.

Once each subject has his or her own sample code, has been recorded in the DNA banking receipt book, and has been accurately linked to a pedigree, the DNA Bank Coordinator creates a file for the subject in which all personal information - including the DNA banking and consent forms and the requisition sheet - can safely be kept. The file is labelled with the person's name, sex, risk status, date of birth, pedigree number, and sample code.

A personal card is also created with all relevant information. These are stored alphabetically in the DNA banking archive. The data on each card are also entered into a file in a custom-made computer programme.

When a subject enters predictive testing and has family members who have previously been banked, a complete pedigree review is performed to ensure the accuracy of all information. The review includes checking against all the cards, files, pedigrees and computer information on all family members.

Laboratory Measures

The blood tubes themselves are transferred to a tube with a label space. In this space is written the subject's full name and his or her subject code. Simultaneously this information is written on the top of the tube, to ensure that tubes cannot be mixed up. After each lab procedure, the DNA Bank Coordinator double-checks that the information on the tube's top corresponds to that on the tube body. Using full names and sample codes on the tubes, rather than simply labeling them anonymously #1 and #2, ensures accuracy in testing.

The DNA in each blood sample is extracted and aliquoted immediately upon arrival. This is to ensure accuracy, to prevent blood sample buildup, and to achieve maximum yield and quality of DNA. At every stage of DNA extraction and storage, all numbers are double-checked.

The success of the above procedures depends, of course, on the accuracy of blood sample labeling in the clinic where the blood is drawn.

Quality Control Procedures for Direct Assessment of CAG Repeat Length

Initially clinical status, date of birth and place in the pedigree are checked. Polymerase chain reaction (PCR) sheets are made up and the sample number is checked against the cards in the DNA bank. Where possible, an affected family member and parents are run simultaneously. All samples from the individual in the DNA bank are run simultaneously to allow assessment for sample mix-up. Where possible, if only an old sample from the bank is available, a new sample is also requested.

Once the PCR run is set up, the samples and the sample numbers are checked against the PCR sheet to make sure that the order is correct and that the actual sample used is correct. The size of the CAG repeat length is checked against an M13 sequencing ladder, and also against samples of CAG sizes of different lengths, which have been cloned and sequenced, and therefore the precise number of CAG repeat lengths of these clones is known. The total CAG length is done twice, using both hot and cold PCR, and CCG length is also assessed as previously described (Goldberg et al. 1993; Andrews et al. 1994). Where other family members are available, including children and parents, they are included for CAG repeat size to ensure there is no sample mix-up.

The reading of the CAG repeat length is done twice, by two different individuals, along with recording and sample codes, which are also checked twice. Each report is written up individually, with date of birth, sample number, predictive testing number, and results checked twice. This report is assessed again prior to signing.

Acceptance of Predictive Testing

Despite the fact that predictive testing has been available since 1986, only approximately 25% of persons eligible for predictive testing by virtue of having a relative affected with this illness have chosen to participate in this programme. Similar low acceptance rates have been reported in other programmes (Crauford et al. 1989; Brandt et al. 1989). In other words, the majority of individuals have at this point chosen not to use predictive testing. Detailed studies on the reasons for the relatively low acceptance rate for predictive testing have revolved around the fact that many persons feel there is not point in undergoing predictive testing if there is no way to modify the course of this illness. Thus, the availability of treatment appears to be a major and significant factor in influencing the acceptibility of predictive testing for late-onset disorders (Wiggins et al. 1992; Babul et al. 1993).

Furthermore, demand for prenatal testing has ever been lower, even if we include those individuals who were clearly informed of this as an option. Only approximately 18% of such eligible and pregnant individuals who were already participating in predictive testing for themselves chose to have prenatal testing (Adam et al. 1993). The most frequently cited reason for not choosing prenatal testing was the belief that a cure would be found in time for their children. In addition, the prospect of terminating a fetus at increased risk for a disorder that would only have its onset in later life was deemed to be unacceptable for many persons.

Therefore it is clear that the acceptance of predictive testing for late-onset illnesses is likely to depend, at least in part, on the burden of this illness in families. As treatment becomes more likely for disorders such as familial hypercholesterolemia or some familial cancer syndromes, predictive testing will be more acceptable. However, for severe illnesses for which there is currently no way to interfere in the progression or the cause of this illness - such as HD, Alzheimer's disease, Prion diseases and hereditary motor sensory neuropathy - it is likely that the majority of persons at-risk will not participate in predictive testing programs.

Relevance to Other Disorders

Currently predictive testing has been primarily performed on disorders where identification of the mutation for the disease leads one to predict that this person will manifest with this illness he or she live long enough. However, in the future it is likely that certain genetic markers that may identify susceptibility to common disorders - such as coronary artery disease, hypertension, diabetes and certain forms of cancer - will become available. This information will in all likelihood be provided by other health care professionals and not geneticists, as there are too few trained geneticists and genetic counsellors to fulfill this demand (Benjamin et al. 1994). There will be significant potential for misunderstanding, particularly in view of the fact that it will be easy to confuse the genetic suspectibility for disease with onset of disease itself. Genetic markers are unlikely themsleves to indicate that a person will definitively develop signs and symptoms of the illness. In reality, this will represent some modification of risk but not a definitive predictor for multi-factorial disease. In many instances, gene-environment modification of the genotype will be a key factor in the pathogenesis of the illness. Therefore, it is most appropriate that tests that are specifically for multifactorial diseases be used with significant caution.

Principles for routine testing for genetic susceptibility can draw significant lessons from testing for HD. In particular, pre-test counselling is absolutely imperative; in addition, it is important for the individuals who are being tested to be fully informed of the risks and benefits and make an autonomous decision with regard to having this test. At the present time, predictive testing for HD, is not offered for children in Canada, as there is not deemed to be any benefit. The principle that testing children for adult-onset illness should only be undertaken when there is some preventative or therapeutic measures exist, needs to be carefully considered.

In addition, issues of confidentiality, particularly in small rural communities, need to be stringently safeguarded. There is presently no justification for population screening for late-onset illnesses; rather, specific family-based testing where there is appropriate family history of certain disorders and where the genetic findings can be interpreted in light of the family history may be appropriate.

In addition, important issues with regard to quality control as outlined for HD are essential for any laboratory doing DNA testing. Quality assurance programmes are necessary as for many of these illnesses the significance of these findings is so great that the goal of no errors in laboratory testing should be the target.

It is also important for genetic counselling to be non-directive, particularly when issues related to procreation are being discussed. In many instances the values of the patients may not be identical to the values of those providing care for these patients, and in these instances directiveness may be damaging. Although non-directive approaches are generally advocated, it is important for the person providing testing to fully explore the implications of these tests and their social context. It should also be recognised that, as new treatments for many late-onset illnesses become available, predictive testing will be more acceptable and will potentially be followed by treatment that may either prevent or ameliorate the progression of the illness.

HD represents the disorder for which predictive testing has now been provided for the longest period. More data are needed concerning the long-term effects of predictive testing in an effort to see whether persons who have had testing still have significant benefits some considerable time after the test. Furthermore, new programmes are currently underway to assess the impact of having persons other than geneticists provide these results in the community. These results will be most important as they will have relevance to the involvement of many other health professionals in providing genetic information to their patients.

Acknowledgements. With thanks to colleagues in the Canadian Collaborative Study of Predictive Testing for Huntington Disease. This work is supported by the Canadian Genome and Technology (CGAT) Program. Michael Hayden is a career investigator of the British Columbia Children's Hospital. I gratefully acknowledge the support of important members of our predictive testing team, including Shelin Adam, Sandy Wiggins, Maurice Bloch, Bill McKellin and Michael Burgess.

References

Adam S, Wiggins S, Whyte P, Bloch M, Shokeir MHK, Soltan H, Meschino W, Summers A, Suchowersky O, Welch JP, Huggins M, Theilmann J, Hayden MR (1993) Five year study of prenatal testing for Huntington disease: Demand, attitudes and psychological assessment. J Med Genet 30: 549-556

Andrew SE, Goldberg YP, Theilmann J, Zeisler J, Hayden MR (1994) A CCG repeat polymorphism adjacent to the CAG repeat in the Huntington disease gene: Implications for diagnostic accuracy and predictive testing. Hum Mol Genet 3: 65-67

Babul R, Adam S, Kremer B, DuFrasne S, Wiggins S, Huggins M, Theilmann J, Block M, Hayder MR (1993) Attitudes toward direct predictive testing for the Huntington disease gene: Relevance for other adult-onset disorders. JAMA 270: 2321-2325

Benjamin CB, Adam S, Wiggins S Theilmann JL, Copley TT, Bloch M, Squitier F, McKellin W, Cox S, Brown SA, Kremer HPM, Burgess M, Meshino W, Summers A, MacGregor D, Buchanan J, Greenberg C, Carson N, Hayden MR (1994) Proceed with care: direct predictive testing for Huntington disease. Am J Hum Genet 55: 606-617

Bloch M, Adam A, Wiggins S, Hayden MR (1992) Predictive testing for Hundington disease. The experience of those receiving an increased risk. Am J Med Gen 42: 499-507

Brandt J, Quaid KA, Folstein SE, Garber P, Maestri NE, Abbott MH, Slavney PR, Franz ML, Kasch L, Kazazian HH Jr (1989) Presymtomatic diagnosis of delayed-onset disease with linked DNA markers: the experience in Huntington's disease. JAMA 261: 3108-3114

Copley T, Wiggins S, Dufrasne S, Bloch M, Adam S, McKellin W, Hayden MR and the Canadian Collaborative Study for Predictive Testing for Huntingon Disease. Are we all of one mind? Clinicians' and patients' opinions regarding the development of a service protocol for predictive testing for Huntington disease. Am J Med Genet, in press

Crauford D, Dodge A, Kerzin-Storrar L, Harris R (1989) Uptake of presymptomatic predictive testing for Huntington's disease. Lancet ii: 603-605

Fox S, Bloch M, Fahy M, Hayden MR (1989) Predictive testing for Huntington disease: I. description of a pilot project in British Columbia. Am J Med Genet 32: 211-216

Goldberg YP, Andrew SE, Clarke LA, Hayden MR (1993) A PCR method for accurate assessment of trinucleotide repeat expansion in Huntington disease. Human Molecular Genetics 2:6: 635-636

Gusella JF, Wexler NS, Conneally PM, Naylor SL, Anderson MA, Tanzi RE, Watkins PL, Ottina K, Wallace MR, Sakaguchi AY, Young AB, Shoulson I, Bonilla E, Martin JB (1983) A polymorphic DNA marker genetically linked to Huntington's disease. Nature 306: 234-238

Huggins M, Bloch M, Wiggins S, Adam S, Suchowersky O, Trew M, Klimek M, Greenberg CR, Eleff M, Thompson LP, Knight J, MacLeod P, Girard K, Theilman J, Hedrick KA, Hayden MR (1992) Predictive testing for Huntington disease in Canada: Adverse effects and unexpected results in those receiving a decreased risk. Am J Med Gen 42: 508–515

Lam RW, Bloch M, Jones BD, Marcus AM, Fox S, Amman W, Hayden MR (1988) Psychiatric morbidity associated with preclinical testing for Huntington disease. J Clin Psych 444–447

Mennie ME, Holloway SM, Brock DJH (1990) Attitudes of general practitioners to presymptomatic testing for Huntington's disease. J Med Genet 27: 224–227

Perry TL (1984) Some ethical problems in Huntington chorea. CMAJ 125: 1098–1100

Schoenfeld M, Myers RH, Cupples A, Berkman B, Sax DS, Clark E (1984) Increased rate of suicide among patients with Huntington's disease. J Neurol Neurosurg Psychiat 47: 1283–1287

The Huntington Disease Collaborative Group (1993) A novel gene containing a trinucleotide repeat that is expanded and unstable on Huntington's disease chromosomes. Cell 72: 971–983

Thies U, Bockel B, Bochdalofsky V (1993) Attitudes of neurologists, psychiatrists, and psychotherapists towards predictive testing for Huntington's disease in Germany. J Med Genet 30: 1023–1027

Thomassen R, Tibben A, Niermeijer MF, Van der Does E, Van de Kamp JJ, Verhage F (1993) Attitudes of Dutch general practitioners towards presymptomatic DNA-testing for Huntington disease. Clin Genet 43: 63–68

Wiggins S, Whyte P, Huggins M, Adam S, Theilmann, J, Bloch M, Sheps SB, Schechter MT, Hayden MR (1992) The psychological consequences of predictive testing for Huntington's disease. New Engl J Med 327: 1401–1405

Molecular Genetics of Mental Disorders: Facts and Hopes

J. MALLET and R. MELONI

The etiology of psychiatric diseases such as manic depressive illness (MDI) and schizophrenia (SCH), the two major psychoses, is not known. Compelling evidence from family, twin and adoption studies demonstrates that genetic factors play a role in the predisposition to these diseases. Other behavioral disorders (for example: alcoholism, Alzheimer's disease, autism, Tourette's syndrome) present a similar genetic predisposition. However, the mode of inheritance of these diseases remains elusive: segregation studies show that they are heterogeneous, poligeneous, present incomplete penetrance and are clearly influenced by environmental factors. In this context, the search for genetic linkage by the lod score method is prone to error because of diagnostic misclassification and misspecification of parameters. Thus, vulnerability genes may be difficult to find even by systematic screening of the entire genome using classical linkage analysis unless a very large panel of reference families is used. Such a study (systematic screening on a very large panel of families) is currently being conducted for MDI and SCH under the patronage of the European Science Foundation (E.S.F.) by several European groups coordinated by our laboratory. This study is a collaborative effort reuniting 20 laboratories from 12 European countries that have gathered a total of around 200 multiplex families with either MDI or SCH for a total of around 2000 subjects of whom 800 are patients. The DNA samples of this population are being studied using the panel of 400 microsatellite markers that cover the whole genome (Weissenbach et al. 1992). This study is conducted utilising high performance robotic devices for the most cumbersome and time consuming tasks at "Genethon." "Genethon" is the scientific branch of the "Association Française contre les Myopathies" a charity foundation sustaining genetic studies in the field of muscular dystrophies and willing to lend its technical prowess to other scientific projects that gather very large samples to be studied.

However, even if classical linkage analysis may detect genes that are centrally involved in the development of a disease (Greenberg 1993), this is inadequate when there are multiple genes acting independently (genetic heterogeneity), additively (polygeneity) of epistatically (Rich 1990). Moreover this method is not appropriate when there are epistatic interactions between genes and

K. Berg, V. Boulyjenkov, Y. Christen (Eds.)
Genetic Approaches to Noncommunicable Diseases

the environment (Cloninger 1994). The genetic complexity of the major psychoses and the constraints of the classical lod score method can thus explain the conflicting results obtained by several groups over the past few years (for review see Owen and Mullan 1990; Pauls 1993). These results have convinced researchers that, although analysis with the classical lod score method may be valuable with large numbers of multiplex families with hereditary MDI or SCH, as was done in the E.S.F. project, the complex nature of these diseases demands also an additional alternative approach. Thus, to complement the lod score parametric method, nonparametric methods are needed such sib-pair (Green and Woodrow 1977) and affected pedigree member (Weeks and Lange 1988) in family studies as well as association analysis (Cooper and Clayton 1988) for the study of sporadic cases and their matched, unaffected controls using candidate genes. The advantage of using the nonparametric approach is that it does not require any assumption about the mode of transmission, thus allowing the identification of susceptibility loci that are neither necessary nor sufficient but only increase the risk for a given disease. It is also possible that genetic variants rather than "yes or no" mutations, determine genetic predisposition to psychiatric disease. Such variants could have varying genetic impact on the phenotype and could also interact with other genetic and/or environmental factors. Many association studies are now underway to detect such partial genetic effects.

As a paradigmatic example of the different strategies employed for a composite approach to the genetics of psychiatric diseases, we report here some results obtained using a candidate gene for MDI. The tyrosine hydroxylase (TH) gene encodes the rate-limiting enzyme in catecholamine synthesis and is thus a candidate gene for psychiatric diseases. The expression of the TH gene is regulated by an elaborate array of biochemical mechanisms in response to environmental factors. Changes in TH activity may result from both short-term (phosphorylation by kinases, Zigmond et al. 1989) and long-term (transynaptic regulation of TH mRNA levels) mechanisms (Icard-Liepkalns et al. 1992).

The human TH gene generates at least four different mRNAs through alternative splicing events. These mRNA species differ in their coding regions and thus lead to the production of enzymes with different characteristics (Le Bourdelès et al. 1991). More than one form appears to be expressed in individual neurons in the brain. The presence in the same tissue of several types of TH mRNA that encode enzymes with different characteristics raises the possibility that alternative splicing plays a role in the regulation of catecholamine biosynthesis in normal and pathological neurons, in addition to transcriptional and posttranslational events (Dumas et al. 1992).

Consequently, the TH gene has been the focus of much attention following an initial report (Egeland et al. 1987) of linkage, in an extended Amish family, between MDI and markers at the Insulin (INS) and H-ras-1 (HRAS) genes which are both linked to the TH gene at the 11p15.5 locus. However, it has not been possible to confirm this finding in other populations or in the extension of the initial Amish pedigree (Kelsoe et al. 1989). Nevertheless, the significance of the initial results has not been completely dismissed because positive linkage has repeatedly been found in the core pedigree. These conflicting results may then confirm that genetic heterogeneity (inter or intrafamilial), polygenicity and interaction with environmental factors play a role in the disease. Furthermore, several groups recently reported linkage data indicating that this locus could not be excluded in several families (Byerley et al. 1992; Lim et al. 1993) consistent with the implication of genetic heterogeneity.

The utilisation of association analysis has also given positive results for the TH locus. We identified an association between MDI and Taq 1 and Bgl II RFLP markers, in the 5′ and 3′ regions of the TH gene respectively (Leboyer et al. 1990). Two other studies have shown a similar frequency of the 3′ allele in individuals with MDI, but failed to obtain a statistically significant result (Körner et al. 1990; Gill et al. 1991). Nevertheless, when all the data for the 3′ allele are pooled, a significant association with MDI ($P < 0.0001$) is found. Further analysis by our laboratory of the TH locus has shown that a polymorphic and highly informative marker of the microsatellite type located in the first intron of the TH gene is associated with MDI (Meloni et al. 1995).

We estimated the allele frequencies using a recently developed method (Boehnke 1991) at TH, INS and H-RAS loci in 11 French pedigrees. Particular alleles, as well as haplotypes sharing these alleles, display higher frequencies than expected and that there is linkage disequilibrium in this region in affected individuals (Mallet et al. 1992, and manuscript submitted). These findings support the conclusion that there is a DNA sequence in the TH-INS-HRAS region that, in concert with other genes and/or environmental factors, causes predisposition to MDI.

All these various findings are thus consistent with the TH gene being involved in the physiopathology of affective disorders. Further studies are under way in our laboratory to identify susceptibility factors to MDI at the TH locus.

The importance of this alternative approach with non parametric methods to the study of psychiatric diseases is also strengthened by the preliminary positive result obtained for linkage of predisposing factors to MDI in chromosome 18 (Berrettini et al. 1994). These observations will undoubtedly lead to the application of a similar approach to re-evaluate previous negative results obtained with the lod score method. This could be the case for the dopamine receptors, which are strong candidate genes for psychiatric diseases and for which there is a frustrating lack of evidence of their implication in SCH.

Moreover a relatively new technique of quantitative trait loci (QTL) mapping allows identification of individual genes that contribute to the development of a trait determined by more than one gene (Lander and Botstein 1989). The application of QTL in animal models could indicate new candidate genes for complex behavioral disorders (Berrettini 1993; Plomin et al. 1994).

The preliminary positive results with alternative approaches to the genetics of psychiatric diseases are also confirming that epistatic and environmental factors can influence the expression of psychiatric diseases (Berrettini et al. 1994). These environmental factors, however, may be difficult to quantify and include phenomena as disparate as developmental events (Cloninger 1994), non-shared environment, and stochastic DNA events, defined as somatic mutations, imprinting and expansible DNA sequences (Plomin et al. 1994).

This complex array of factors putatively involved in the predisposition to psychiatric disease, encourages renewed consideration of the candidate genes and to the types of genetic information they could reveal. Direct searching for susceptibility mutations (Cotton and Malcom 1991; Cotton 1993; Grompe 1993) in candidate genes may be then a more effective approach than allelic association for progress in psychiatric genetics. The detection of subtle genetic variants with effects on the phenotype is more probable than the detection of clear cut mutations. This reasoning has led us to start a systematic screening of the TH gene for genetic variants that could be implicated in the pathology of MDI. Since the expression of the TH involves complex and multiple regulatory systems, we are scanning both the coding region of the gene and also the proximal promoter region and the first intron for sequence variations by Single Strand Conformation Polymorphism (SSCP) analysis and direct sequencing. SSCP analysis is a recently developed method for detection of polymorphisms in the DNA that does not require the presence of a restriction site (Orita et al. 1989a,b). The migration of single-strand nucleic acid fragments in non-denaturing polyacrylamide gel is dependent on both their size and shape. The electrophoretic mobility of the fragment thus depends not only on its length but also on its folded structure that is determined by intramolecular interactions and, therefore, by its sequence. Thus, in SSCP analysis, a mutant sequence is detected as a change of electrophoretic mobility caused by its altered folded structure. SSCP can distinguish most conformational changes caused by differences as small as one base substitution in a several-hundred-base-long fragment, although it is not possible to predict "a priori" the shift of electrophoretic mobility induced by the mutation (Sheffield et al. 1993).

The identification of sequence variations in candidate genes and their relationship to the phenotype will help better define the different clinical entities in psychiatric genetics. This, is turn, will contribute to improve further our understanding of the genetic factors implicated in mental disease and evaluate their interaction with environmental factors. The rapid advances in automated processes and the development of novel statistical methods are keys to the analysis of the genetic basis of major psychiatric and behavioral disorders, the understanding of which is improving and is bound to progress at a rapid pace in the near future.

References

Berrettini WH (1993) Quantitative trait loci mapping: a novel approach for candidate genes for psychiatric diseases. Psych Gen 3: 203–205

Berrettini WH, Ferraro TN, Goldin LR, Weeks DE, Detera-Wadleigh S, Nurnberger JI Jr, Gershon ES (1994) Chromosome 18 DNA markers and manic-depressive illness: evidence for a susceptibility gene. Proc Nat Acad Sci USA 91: 5918–5921

Boehnke M (1991) Allele frequency estimation from data on relatives. Am J Human Genet 48: 22–25

Byerley W, Plaetke M, Hoff M, Jensen S, Holik J, Reimherr F, Mellon C, Wender P, O'Connel P, Leppert M (1992) Tyrosine hydroxylase gene not linked to manic-depression in seven of eight pedigrees. Human Heredity 2: 259–263

Cloninger CL (1994) Turning point in the design of linkage studies in schizophrenia. Am J Med Gen (Neuropsy Gen) 54: 83–92

Cooper DN, Clayton JE (1988) DNA polymorphism and the study of disease association. Human Gen 78: 299–312

Cotton RGH (1993) Current methods of mutation detection. Mutation Res 285: 125–144

Cotton RGH, Malcom ADB (1991) Mutation detection. Nature 353: 582–583

Dumas S, Horellou P, Helin C, Mallet J (1992) Co-expression of tyrosine hydroxylase messenger RNA 1 and 2 in human ventral mesencephalon revealed by digoxigenin- and biotin-labelled oligodeoxyribonucleotides. J Chem Neuroanat 5: 11–18

Egeland JA, Gerhard DS, Pauls DL, Sussex JN, Kidd KK, Allen CR, Hostetter AM, Housman DE (1987) Bipolar affective disorders linked to DNA markers on chromosome 11. Nature 325: 783–787

Gill M, Castle D, Hunt M, Clemens A, Sham P, Murray RM (1991) Tyrosine hydroxylase polymorphisms and bipolar affective disorder. J Psych Res 25: 179–184

Green JR, Woodrow JC (1977) Sibling method for detecting HLA-linked genes in diseases. Tissue Antigens 9: 31–35

Greenberg DA (1993) Linkage analysis of "necessary" disease loci versus "susceptibility" loci. Am J Human Gen 52: 135–143

Grompe M (1993) The rapid detection of unknown mutations in nucleic acids. Nature Genet 5: 111–112

Icard-Liepkalns C, Faucon Biguet N, Vyas S, Robert JJ, Sassone-Corsi P, Mallet J (1992) AP-1 Complex and c-Fos transcription are involved in TPA provoked and trans-synaptic inductions of the tyrosine hydroxylase gene: insights into long-term regulatory mechanisms. J Neurosci Res 32: 290–298

Kelsoe JR, Ginns EI, Egeland JA, Gerhard DS, Goldstein AM, Bale SJ, Pauls DL, Long RT, Kidd JJ, Conte G, Housman DE, Paul SM (1989) Re-evaluation of the linkage relationship between chromosome 11p loci and the gene for bipolar affective disorder in the old order Amish. Nature 342: 238–243

Körner J, Fritze J, Propping P (1990) Alleles at the tyrosine hydroxylase locus: no association with manic-depressive illness. Psychi Res 32: 275–280

Lander ES, Botstein D (1989) Mapping Mendelian factors underlying quantitative traits using RFLP linkage maps. Genetics 121: 185–199

Le Bourdelès B, Horellou P, Le Caer JP, Denefle P, Latta M, Haavik J, Guibert B, Mayaux JF, Mallet J (1991) Phosphorylation of human recombinant tyrosine hydroxylase isoforms 1 and 2: an alternative phosphorylated residue in isoform 2, generated through alternative splicing. J Biol Chem 266: 17124–17130

Leboyer M, Malafosse A, Boularand S, Campion D, Gheysen F, Samolyk D, Henriksson B, Denise E, des Lauriers A, Lepine J-P, Zarifian E, Clerget-Darpoux F, Mallet J (1990) Tyrosine hydroxylase polymorphisms associated with manic-depressive illness. Lancet 335: 1219

Lim L, Gurling H, Curtis D, Brynjolfsson J, Petursson H, Gill M (1993) Linkage between tyrosine hydroxylase gene and affective disorder cannot be excluded in two of six pedigrees. American J Med Gen (Neuropsychiatric Genetics) 48: 223–228

Mallet J, Malafosse A, Leboyer M, d' Amato T, Amadéo S. Abbar M, Gheysen F, Granger B, Henriksson B, Loo H, Poirier MF, Sabaté O, Samolyk D, Zarifian E, Clerget-Darpoux F (1992) Family studies of effective disorders with 11p15.5 DNA markers. In: Mendlewicz J, Hippius H (eds) Genetic Research in Psychiatry. Springer-Verlag, Berlin, pp 164–172

Meloni R, Leboyer M, Bellivier F, Barbe B, Samolyk D, Allilaire JF, Mallet J (1995) Association of manic-depressive illness with a tyrosine hydroxylase microsatellite marker. Lancet 345: 932

Orita M, Hiwahana H, Kanazawa H, Hayashi K, Sekiya T (1989a) Detection of polymorphism of human DNA by gel electrophoresis as single-strand conformation polymorphisms. Proc Nat Acad Sci USA 86: 2766–2770

Orita M, Suzuki Y, Sekiya T, Hayashi K (1989b) Rapid and sensitive detection of point mutations and DNA polymorphisms using the polymerase chain reactions. Genomics 5: 874–879

Owen M, Mullan M (1990) Molecular genetic studies of manic-depression and schizophrenia. Trends Neurosci 13: 29–31

Pauls D (1993) Behavioural disorders: lessons in linkage. Nature Gen 3: 4–5

Plomin R, Owen MJ, McGuffin P (1994) The genetic basic of complex human behavior. Science 264: 1733–1739

Rich N (1990) Linkage strategies for genetically complex traits. I multilocus models. Am J Human Gen 46: 222–228

Sheffield VC, Beck JS, Kwitek AE, Sandstrom DW, Stone EM (1993) The sensitivity of single-strand conformation polymorphism analysis for the detection of single base substitutions. Genomics 16: 325–332

Weeks DE, Lange K (1998) The affected-pedigree-menber method of linkage analysis. Am J Human Gen 42: 315–326

Weissenbach J, Gyapay G, Dib C, Vignal A, Morissette J, Millasseau P, Vaysseix G, Lathrop M (1992) A second-generation linkage map of the human genome. Nature 359: 794–801

Zigmond RE, Schwarzschild MA, Rittenhouse AR (1989) Acute regulation of tyrosine hydroxylase by nerve activity and by neurotransmitters via phosphorylation. Ann Rev Neurosci 12: 415–461

Lp(a) Genes, Other Genes, and Coronary Heart Disease

K. Berg

Introduction

Environmental as well as genetic factors contribute to the etiology of coronary heart disease (CHD) and other atherosclerotic or thrombotic diseases. Life style or dietary factors apparently cause disease preferentially in those with a genetic predisposition. Although the role of genetic factors may be particularly important in cases where disease occurs at a relatively young age, some genetic factors also appear to be of importance for a disease manifested after the age of 60.

Several risk factors or protective factors with respect to CHD, including total cholesterol, low density lipoprotein (LDL) cholesterol (LDLC), high density lipoprotein (HDL) cholesterol (HDLC), apolipoprotein B (apoB), apolipoprotein A-I (apoA-I), and blood pressure, exhibit an impressive level of heritability, typically around 0.5. Increased levels of fibrinogen or homocysteine are also significant risk factors that are genetically influenced, but their level of heritability is more modest. The concentration of Lp(a) lipoprotein is almost exclusively determined by genes (heritability is close to unity). The knowledge that genes contribute to the susceptible state with respect to CHD makes it important to identify genes of major significance so that people at particularly high risk may avail themselves of any known preventive measure, in early adult life.

The Functional Candidate Gene Approach

At present, most efforts to identify genes of importance for CHD or its risk factors follow the *functional candidate gene approach*. A *functional candidate gene* is a gene whose protein product is, or could be, involved in lipoprotein metabolism, lipoprotein structure, thrombogenesis, thrombolysis, fibrinolysis, regulation of blood flow in coronary arteries, regulation of blood pressure, reverse cholesterol transport, regulation of growth of atherosclerotic lesions, or early development of coronary arteries, or is present in atherosclerotic lesions.

K. Berg, V. Boulyjenkov, Y. Christen (Eds.)
Genetic Approaches to Noncommunicable Diseases

The efforts to map the human genome have led to one additonal use of the term "candidate gene." Gene mappers may use this term for genes that are in an area of the genome where a given gene (for example, one causing an inherited disorder) must be located (this knowledge may have emerged from genetic marker studies in families). Genes that are candidate genes because of their position in the genome may be referred to as *positional candidate genes* to distinguish them from *functional candidate genes.*

The functional candidate gene approach was employed long before the term "candidate gene" was coined. The discovery that a genetically determined high level of Lp(a) lipoprotein (Berg 1963) is a risk factor for CHD (Berg et al. 1974) and the demonstration of definite effects on lipid levels of the Ag allotypic polymorphism of LDL (Berg et al. 1976) appear to be the first examples of successful uses of the *functional candidate gene approach.*

Effect of Normal Genes at the Apolipoprotein B Locus on Lipid Levels

Numerous studies have been published to uncover associations between DNA markers at apolipoprotein loci and CHD risk factors or overt disease. Not all of the reported associations have been readily confirmed. One association that appears to stand the test of time is between a restriction site polymorphism at the apoB locus corresponding to amino acid 2488 in the mature protein. Law et al. (1986) observed that the absence of a XbaI restriction site (reflecting a third-base mutation in the codon) was associated with reduced cholesterol and apoB levels, and we readily confirmed this association (Berg 1986). Although some studies have failed to detect an association between this XbaI polymorphism and lipid or apolipoprotein levels, the total amount of information clearly favors this association reflecting a true biological phenomenon. The association is probably caused by linkage disequilibrium between the polymorphism and functionally important domains of the gene.

A recent case/control study (Bφhn and Berg 1994) revealed an association between premature myocardial infarction (MI) and homozygous absence of this XbaI restriction site. The association ($P = 0.007$), which could only be detected in multivariate logistic regression analysis, was not mediated through lipid levels (the disease association was with the XbaI genotype exhibiting the *lowest* cholesterol levels). One may speculate whether a structural change in apoB (in an area in linkage disequilibrium with the XbaI polymorphism) makes the LDL particles more prone to become trapped in arterial walls or increases their atherogenicity for other reasons.

Effects on Risk Factor Level of Genes at Loci Other than Apolipoprotein Loci

Following up on early studies in our laboratory, Pedersen and Berg (1989) found an effect of normal genes at the LDL receptor (LDLR) locus on the population variation in total and LDL cholesterol. This observation was soon independently confirmed, proving the validity of our early postulate that the biological importance of genetic variation at the LDLR locus is not limited to problems relating to classical, autosomal dominant hypercholesterolemia.

Cholesteryl ester transfer protein (CETP) appears to play a vital role in reverse cholesterol transport, the least understood part of lipid metabolism. We observed significant differences between people of different genotypes in a normal DNA polymorphism at the CETP locus with respect to HDLC and apoA-I levels (Kondo et al. 1989). This observation has also been independently confirmed. An association has been reported between an insertion/deletion (I/D) polymorphism at the locus for angiotensinogen I converting enzyme (ACE) and MI (Cambien et al. 1992). Although we were unable to confirm this association in a series of Norwegian CHD patients and controls, we found that parental MI was more frequent in homozygotes or heterozygotes for the D allele than in homozygotes for the I allele (Berge et al. 1994). This confirmation of reports by French workers supports the view that ACE polymorphisms could be important markers of MI risk.

The First Example of Gene-Gene Interaction Affecting CHD Risk Factors

Studying DNA polymorphisms at the LDLR locus in people who had previously been analyzed with respect to the normal polymorphism in apolipoprotein E (apoE), Pedersen and Berg (1989) observed that the well-established cholesterol-increasing effect of the apoE4 gene was present only in people who lacked a normal gene at the LDLR locus identifiable as the presence of a BglI restriction site. Thus, the presence of this normal gene totally obliterated the effect of the apoE4 gene on total and LDL cholesterol. Although the molecular mechanism(s) underlying the phenomenon remains obscure, Pedersen and Berg (1989) concluded that the first example of the effect on a CHD risk factor of interaction between genes at entirely different loci had been detected.

Effect of Smoking on a Genetic Association

The persons included in our study (see above) of the effect of a normal CETP polymorphism on HDLC and apoA-I levels had been questioned with respect to cigarette smoking. When the sample was analyzed with respect to smoking status, it turned out that the effect of this normal CETP polymorphism on

HDLC, as well as apoA-I, was present only in non-smokers (Kondo et al. 1989). Apparently, smokers deprived themselves of the advantage of having as high an HDL level as their genotypes would permit. This finding illustrates the necessity of taking into consideration factors such as smoking, sex and age in studies on risk factor associations.

The Variability Gene Concept

I have suggested (for review, see Berg 1994a) that lipid *variability* may by itself be of clinical importance. It has been known for many years that there are strain differences in several animal species in response to dietary lipid intake. These differences are almost certainly of a genetic nature. Importantly, there is now evidence of significant differences between people with respect to response to changes in lipid intake, suggesting that "high responders" and "low responders" may exist in man as well. The potential importance of lipid *variability* (in addition to absolute lipid levels) in the etiology of CHD, as well as the fact that man is postprandial most of the day, suggested the need for more dynamic approaches to the study of genetic markers and disease risk than the traditional association studies based on measurement of lipids in the fasting state. The method we developed to study the possible effect of genes on risk factor *variability* utilizes the genetic identity of monozygotic (MZ) twins (for details of the method, see Berg 1994a). The method appears to be a valid instrument to study gene-environment interactions and to detect "variability genes." In addition to our own evidence from the study of MZ twins (Berg 1994a), indicating that variability genes exist at the apoB locus with respect to apoB concentration and at the apoA-I locus with respect to cholesterol, papers from other laboratories confirming the existence of variability genes have started to appear. The variability gene concept could be important for the understanding of genetic predisposition to atherosclerotic disease, since an individual's total genetic risk of CHD or other atherosclerotic diseases may depend on his or her *combination* of *level genes* and *variability genes.*

The Lp(a) Lipoprotein

The Lp(a) lipoprotein (Berg 1963) was detected by the use of an animal immunization and antibody absorption strategy aimed at uncovering lipoprotein differences between individuals. Lp(a) lipoprotein was found to be a distinct serum lipoprotein particle, having unique antigenic structures as well as antigens shared with LDL. The long Lp(a) polypeptide chain (or apolipoprotein Lp(a), or apolipoprotein(a)) is attached to apoB in a LDL-like particle. The successful cloning (McLean et al. 1987) of cDNA representing the gene for the Lp(a) polypeptide chain (the LPA gene) revealed striking homology to plasminogen, a much smaller protein important for fibrinolytic/thrombolytic processes. Extreme variability in Lp(a) lipoprotein concentration between individuals and very strict genetic control within the individual are striking characteristics of the Lp(a) lipoprotein. Single gene control of Lp(a) lipoprotein was postulated in the very earliest publications and has since been repeatedly confirmed (for review, see Berg 1994b).

The discovery (Berg et al. 1974) of an association between Lp(a) lipoprotein and CHD has been confirmed in numerous studies. In one study of patients and controls, samples from all participants were examined blindly in two of the most experienced laboratories. There were numerous split samples unknown to the investigators and there was excellent agreement between split samples and laboratories (Rhoads et al. 1986). On the basis of this study, a population-attributable risk of 28% to contract myocardial infarction prior to age 60 was calculated for men having an Lp(a) lipoprotein concentration in the top quartile of the population distribution. There was also an effect of high Lp(a) lipoprotein level in men aged 60 to 69 (Rhoads et al. 1986). Correlation analyses have shown that the level of Lp(a) lipoprotein is essentially independent of other known risk factors or protective factors, and the importance of a high Lp(a) lipoprotein level as a CHD risk factor has been confirmed in prospective studies (Rosengren et al. 1990; Schaefer et al. 1994). Occasional studies that failed to detect the connection between Lp(a) lipoprotein and CHD may have been based on inadequate techniques or reagents (commercially available test kits; Berg 1994c).

Lp(a) lipoprotein contributes to the atherogenic process since it is present in atherosclerotic lesions, but there is also some information from *in vitro* studies tending to confirm the notion that Lp(a) lipoprotein may interfere with thrombolytic/fibrinolytic processes.

The question of the physiological role of Lp(a) lipoprotein is unresolved. In a study of elderly persons we found a much lower proportion of people with a very high Lp(a) lipoprotein level than in the general population. The corresponding excess of people was, however, not found in the lowest quartile of Lp(a) lipoprotein concentrations, but in the second to lowest quartile, perhaps indicating that there is some advantage to having a moderate level of Lp(a) lipoprotein rather than a very low level. This may suggest that genes determining a moderate level of Lp(a) lipoprotein are "longevity genes."

We have recently studied a woman with severe thrombotic disease at age 43, who had given birth to three children with very low birth weights. The placentas had been small and ischemic. Her Lp(a) lipoprotein concentration was above the 99th percentile in the general population. This observation could suggest than an exceptionally high Lp(a) lipoprotein level may interfere with placental circulation and cause poor placental development and function.

Concluding Remarks

The *functional candidate gene* approach is the method of choice in attempts to identify genes contributing to atherosclerosis risk and led to the detection of

Lp(a) lipoprotein as a major genetic risk factor many years ago. Alleles at the LPA locus may contribute to atherogenesis, thrombogenesis, placental ischemia or longevity. Lp(a) lipoprotein, the LPA gene, disease associations of markers at candidate loci that are not mediated through levels of traditional risk factors, and the variability gene concept are major challenges to workers interested in the genetics of atherosclerosis.

Acknowledgements. Work in the authors laboratory was supported by grants from the Norwegian Council on Cardiovascular Disease and Anders Jahres Foundation for the Promotion of Science.

References

Berg K (1963) A new serum type system in man – the Lp system. Acta Pathol Microbiol Scand 59: 369–382
Berg K (1986) DNA polymorphism at the apolipoprotein B locus is associated with lipoprotein level. Clin Genet 30: 515–520
Berg K (1994a) Gene-environment interaction: variability gene concept. In: Goldbourt U, de Faire U, Berg K (eds) Genetic factors in coronary heart disease. Kluwer Academic Publishers, Dordrecht Boston London, pp 373–383
Berg K (1994b) Lp(a) lipoprotein: An overview. Chem. Physics Lipids 67/68: 9–16
Berg K (1994c) Confounding results of Lp(a) lipoprotein measurements with some test kits. Clin Genet 46: 57–62
Berg K, Dahle'n G, Frick MH (1974) Lp(a) lipoprotein and pre-β_1-lipoprotein in patients with coronary heart disease. Clin Genet 6: 230–235
Berg K, Hames C, Dahle'n G, Frick MH, Krishan I (1976) Genetic variation in serum low density lipoproteins and lipid levels in man. Proc Natl Acad Sci (USA) 73: 937–940
Berge KE, Bøhn M, Berg K (1994) DNA polymorphism at the locus for aniotensinogen I-converting enzyme in Norwegian patients with myocardial infarction and controls. Clin Genet 46: 102–104
Bøhn M , Berg K (1994) The XbaI polymorphism at the apolipoprotein B locus and risk of atherosclerotic disease. Clin Genet 46: 77–79
Cambien F, Poirier O, Lecerf L, Evans A, Cambou J-P, Arveiler D, Luc G, Bard J-M, Bara L, Ricard S, Tiret L, Amouyel P, Alhenc-Gelas F, Soubrier F (1992) Deletion polymorphism in the gene for angiotensin-converting enzyme is a potent risk factor for myocardial infarction. Nature 359: 641–644
Kondo I, Berg K, Drayna D, Lawn R (1989) DNA polymorphism at the locus for human cholesteryl ester transfer protein (CETP) is associated with high density lipoprotein cholesterol and apolipoprotein levels. Clin Genet 35: 49–56
Law A, Powell LM, Brunt H, Knott T, Altman DG, Rajput J , Wallis SC, Pease RJ, Priestley LM, Scott J, Miller GJ, Miller NE (1986) Common DNA polymorphism within the coding sequence of the apolipoprotein B gene associated with altered lipid levels. Lancet i: 1301–1303
McLean JW, Tomlinson JE, Kuang W-J, Eaton DL, Chen EY, Fless GM, Scanu AM, Lawn RM (1987) cDNA sequence of human apolipoprotein (a) is homologous to plasminogen. Nature 330: 132–137
Pedersen J, Berg K (1989) Interaction between low density lipoprotein receptor (LDLR) and apolipoprotein E (apoE) alleles contributes to normal variation in lipid levels. Clin Genet 35: 331–337

Rhoads GG, Dahlen G, Berg K, Morton NE, Dannenberg AL (1986) Lp(a) lipoprotein as a risk factor for myocardial infarction. JAMA 256: 2540–2544

Rosengren A, Wilhelmsen L, Eriksson E, Risberg B, Wedel H (1990) Lipoprotein (a) and coronary heart disease: a prospective case-control study in a general population sample of middle aged men. Br Med J 301: 1248–1251

Schaefer EJ, Lamon-Fava S, Jenner JL, McNamara JR, Ordovas JM, Davis CE, Abolafia JM, Lipoel K, Levy RI (1994) Lipoprotein (a) levels and risk of coronary heart disease in men. The Lipid Research Clinics Coronary Primary Prevention Trial. JAMA 271: 999–1003

MED-PED: An Integrated Genetic Strategy for Preventing Early Deaths

R.R. WILLIAMS, I. HAMILTON-CRAIG, G.M. KOSTNER, R.A. HEGELE, M.R. HAYDEN, S.N. PIMSTONE, O. FAERGEMAN, H. SCHUSTER, E. STEINHAGEN-THIESSEN, U. BEISIEGEL, C. KELLER, A.E. CZEIZEL, E. LEITERSDORF, J.C. KASTELEIN, J.J.P. DEFESCHE, L. OSE, T.P. LEREN, H.C. SEFTEL, F.J. RAAL, A.D. MARAIS, M. ERIKSSON, U. KELLER, A.R. MISEREZ, T. JECK, D.J. BETTERRIDGE, S.E. HUMPHRIES, I.N.M. DAY, P.O. KWITEROVICH, R.S. LEES, E. STEIN, R. ILLINGWORTH, J. KANE, and V. BOULYJENKOV

Abstract

New opportunities are emerging for preventing the consequences of serious diseases because of the discovery of causal mutations in specific gene loci. MED PED is an international collaboration to *M*ake *E*arly *D*iagnoses and *P*revent *E*arly *D*eaths in *MED*ical *PED*igres.

The initial target of this effort is heterozygous familial hypercholesterolemia (FH). This is one of the most common and best understood serious single gene disorders. Clinical and genetic diagnoses are reliable, relatively inexpensive, and available worldwide. Potent and effective medications are available that act specifically at the site of genetically defective low density lipoprotein receptors. Clinical trials have established the ability of medical therapy to normalize cholesterol levels, arrest or reverse atherosclerotic coronary artery lesions, and significantly reduce morbidity and mortality.

Over 10 million persons in the world have FH and about 200,000 of them die with premature ischemic heart disease each year. Unfortunately few patients with FH are receiving the benefit of recent advances in the diagnosis and treatment of FH. Pooled estimates from 14 countries indicate that 80% of the patients with FH are *not diagnosed*, 84% are *taking no medication* to lower their cholesterol, and only 7% have reasonably well treated cholesterol levels. Moreover, few are adequately evaluated to detect and treat already established coronary artery disease.

The international MED-PED collaborators have organized to share successful approaches and combine efforts to find and help persons with FH. The initial computer registries in 14 countries contain 13,118 FH patients together with their relatives and physicians. Four international MED-PED subcommittees have been organized to focus on specific efforts:

1) Patient and Physician Education: to share educational materials and support programs

K. Berg, V. Boulyjenkov, Y. Christen (Eds.)
Genetic Approaches to Noncommunicable Diseases

2) Government Affairs and Publicity: to increase public awareness and funding
3) Research and Molecular Genetics: to promote collaboration & research progress
4) Computer Data Management: to standardize tools for FH family registry data

A major feature of the MED-PED approach is concentration on high risk families for rapid case finding. From each known index case in the registry, several new FH cases can be identified among close relatives; sometimes many FH cases can also be found among distant relatives identified using pedigree expansion. Education, treatment, and long term support are also thought to be more effective in families. The MED-PED approach has been tested in some locations for up to six years with considerable success. The MED-PED model is also appropriate for other treatable dominant single gene diseases such as familial defective Apo B, dominant Type III hyperlipidemia, glucocortoid remediable aldosteronism (GRA) with hypertension and early strokes, long QT sudden arrhythmic death syndrome, and some forms of breast and colon cancer.

Need for Practical New Approaches to Follow Gene Discovery

New discoveries in molecular genetics capture the headlines of newspapers on a regular basis. Approximately three billion dollars are being spent worldwide to "map the human genome". What do we hope to gain from this massive investment in molecular genetics? Most of us hope it will lead to a better understanding of disease causation, to more effective detection and treatment of genetically promoted diseases, and to better prevention of serious disease consequences.

Chronic diseases like early familial coronary heart disease and underlying genetic causes like heterozygous familial hypercholesterolemia (FH) present a new challenge for those interested in applying genetic principles to prevent disease consequences. For many years, neonatal screening followed by early dietary treatment have helped families with an inherited error of metabolism, phenylketonuria (PKU). In many locations with well-organized prenatal screening programs, infants with PKU rarely (if ever) escape diagnosis and prompt attention from vigilant medical professionals. In contrast, genes like FH can act like a time bomb, often ticking too softly to be heard from birth until sudden death from myocardial infarction at age 45 in an asymptomatic male or 10-15 years later in an unsuspecting, otherwise healthy, female.

Now that single gene diseases with delayed manifestation in adults like FH are just as treatable as PKU, practical approaches need to be developed to promote comparable levels of diagnosis and treatment for disorders like FH. We present here the experiences of MED-PED collaborators in 14 countries in finding and helping families with FH.

Heterozygous FH, a Best Case Scenario of a Treatable Genetic Disease

FH is a dominantly inherited defect in low density lipoprotein (LDL) receptors carried on the short arm of chromosome 19. With one normal gene and one defective gene at the LDL receptor locus, single gene carriers have only half the normal number of receptors in the liver that recognize and remove circulating LDL-cholesterol from the bloodstream. As a result, these heterozygotes typically have twice the normal LDL-cholesterol levels by age two and, without treatment, for the rest of their shortened lives (Yamamoto et al. 1989).

FH is one of the most common single gene determinants of severe cardiovascular disease. In most countries it is estimated to occur once in every 500 persons, totalling about 10.7 million in the world. Founder effects have led to much higher rates of FH in some populations such as the French Canadians and Afrikaners. An estimate of the number of FH persons in collaborating MED-PED countries is presented in Table 1. In 1985 Brown and Goldstein received the Nobel Prize for their work discovering the LDL receptor defects that cause FH. To date about 150 different causal mutations at the LDL receptor locus have been reported (Hobbs et al. 1992), and about 10 more are known to the MED-PED collaborators. The last column in Table 1 indicates how many of the 160 mutations have been detected in one or more patients from each of the 14 countries participating in the MED-PED FH network.

The methods for diagnosing FH are well developed. In many locations DNA testing is available for accurate diagnosis of locally common mutations. In addition, reliable clinical criteria using inexpensive and widely available blood lipid testing in index cases and relatives have been published showing 98% specificity compared to DNA testing (Williams et al. 1993a). Sensitivity for clinically diagnosing FH was 87% in first degree relatives. Different cholesterol cut points have been established to diagnose FH in general population screening and FH family screening. For example, 310 mg/dl is too low for diagnosing FH in

Table 1. Prevalence of heterozygous familial hypercholesterolemia

Country	Population	Number with FH	Number of mutations
Australia	17,800,000	35,600	?
Austria	7,500,000	15,000	?
Canada	29,000,000	65,000	15 +
Denmark	5,000,000	10,000	25
Germany	80,000,000	160,000	22
Hungary	10,500,000	21,000	3
Israel	5,400,000	10,800	10
Netherlands	15,000,000	38,000	30
Norway	4,000,000	13,000	20
South Africa	40,000,000	78,000	15
Sweden	8,500,000	17,000	10
Switzerland	7,000,000	15,000	2
United Kingdom	55,000,000	110,000	50
United States	251,000,000	502,000	28
Total MED-PED countries	535,700,000	1,090,400	160
World population	5,363,000,000	10,700,000	160

the general population (only 5% above 310 have FH), but in first degree relatives of FH probands 95% of those above 310 have FH. This approach recognizes that the probability of FH at a given cholesterol level depends highly on the probability of having FH before testing. The criteria for the diagnosis of FH by age and relationship to a known FH case are given in Table 2.

Because FH is a dominant gene, 50% of first degree relatives will also be affected (parents, offspring, and siblings) and *on the affected side of the family*, 50% of second degree relatives on average will also have FH (aunts, uncles, nieces, nephews) as well as 25% of first cousins. Screening of all of

Table 2. Validated criteria for the diagnosis of heterozygous familial hypercholesterolemia

1. DNA diagnosis of an accepted functional LDL-receptor mutation[a].
2. DNA diagnosis from unambigous genetic linkage to the LDL receptor locus for severely hypercholesterolemic relatives in a given family.
3. Without a gene marker, use these clinical criteria[b] for the INDEX CASE:

(All 5 criteria must be met)	*age 40*	*age 30*	*age 20*	*under 18*
a. VERY high total Cholesterol[c]:	>360/9,3	>340/8,8	>290/7,0	>270/7.0
b. Normal triglycerides[c]:	<200/2,3	<180/2,0	<150/1,7	<100/1,1
or very high LDL Cholesterol[c]	>260/6,7	>240/6,2	>210/5,4	>200/5,2

c. No secondary cause of high cholesterol (nephrotic syndrome, pregnancy, etc.)
d. At least one pediatric relative (child, grandchild, niece or nephew, etc. < 18 years of age with very high cholesterol (total cholesterol > 270) OR at least one close relative with tendon xanthomas.
e. Dominant expression in the family :

··· about half of siblings and offspring affected.
··· bimodal (clear separation between normal and abnormal in relatives)

4. Among CLOSE RELATIVES of CONFIRMED FH index cases use these criteria:

(Both criteria must be met)	*age 40*	*age 30*	*age 20*	*under 18*
a. High total cholesterol[c]	>300/7,8	>280/7,2	>240/6,2	>220/5,7
or High LDL cholesterol[c]	>215/5,6	>195/5,0	>175/4,5	>165/4,3

b. No secondary cause of high cholesterol (nephrotic syndrome, pregnancy, etc.)

[a] Some functional mutations of the apolipoprotein B locus (eg. apo B 3500) have similar consequences on LDL cholesterol and early coronary heart disease. These are often referred to as "familial defective apo B" or "FDB". Some MED-PED registries are also including persons and families with FDB in their efforts to find and help person with FH.

[b] These clinical criteria were validated using DNA diagnoses and shown to have 98% specificity and 87% sensitivity in relatives in FH families as reported in Am J Cardiol 1993; 72: 171-176. *The cholesterol criteria for diagnosing FH as presented in this table were derived from the US LRC population data as noted in this publication. In some countries, higher population mean cholesterol levels, will require raising the cholesterol levels for diagnosing FH in order to preserve the same level of specificity, and will also result in a lower sensitivity, increasing the need for DNA testing to diagnose FH.*

[c] Cholesterol and triglyceride levels are given first in mg/dl followed by mmol/ml.

these close relatives in a family with FH should be a high priority because they have such a high risk of a treatable fatal disease.

FH causes coronary atherosclerosis with death from myocardial infarction in the 20s in some male and female FH heterozygotes, and typically in men ages 35–55 and women 10–15 years older. These early deaths can be prevented! HMG Co-A Reductase inhibitors (nicknamed HMGs) were introduced a few years ago and have proven to be true "miracle drugs" for FH. HMGs work by upregulating LDL receptors, helping persons with FH to compensate for the half that are detective by getting better productivity out of the half that are normal. FH heterozygotes taking HMGs in combination with a very strict diet and/or other agents such as resin binders and nicotinic acid can often achieve normal or near normal cholesterol levels (Illingworth 1993), even though some of them started out with very high levels (above 400 mg/dl or 10 mmol/l). In a study of such effective cholesterol lowering therapy for three years, coronary angiograms before and after therapy showed significant delay or even regression of atherosclerotic lesions in men and women with FH (Kane et al. 1990). Recently a placebo-controlled trial of simvastatin in 4,444 scandinavian men and women with prior coronary disease showed a 40% reduction in coronary mortality and a 30% reduction in total mortality (Scandinavian Simvastatin Survival Study Group 1994). These results offer considerable hope to FH heterozygotes, many of whom have coronary risk similar to the patients in this "4S Study."

Some question the "cost effectiveness" of drug therapy for elevated cholesterol, especially for "primary prevention" (i.e. before the first heart attack). However in persons with FH, primary prevention of coronary disease by treating cholesterol levels is reported to save money as well as lives in contrast to the controversial expense versus benefit of primary prevention treatment of cholesterol in many other settings (Goldman et al. 1993).

In summary, compared to other single gene diseases, FH represents a "best case scenario" for developing practical approaches to help affected patients and their relatives for several reasons:

1) a large number of casual mutations (160) have been identified,
2) the pathophysiological mechanism is well understood (defective LDL receptors in the liver remove only half the normal amount of cholesterol from the blood),
3) effective treatment is widely available (potent medications with particularly beneficial effects on LDL receptors can now help normalize cholesterol levels in heterozygous FH patients),
4) evidence shows that treatment helps halt or reverse the disease process in FH patients (atherosclerotic lesions in coronary arteries) and reduces mortality rates in other high risk patients with elevated cholesterol levels,
5) cost/benefit analyses report drug treatment of FH to be cost effective,
6) high cholesterol and heart attacks are probably among the least sensitive illnesses to discuss and trace in families.

Why Do FH Heterozygotes Continue to have Preventable Early Deaths?

As shown in Table 3, in data combined from the 14 MED-PED countries only about 20% of persons with FH are estimated to be diagnosed. Furthermore even fewer are estimated to be taking *any* cholesterol lowering medication. Very few are receiving the life-saving benefit of good control of cholesterol levels. It has been almost 10 years since Brown and Goldstein were given the Nobel Prize for discovering the defective genetic mechanism causing FH. It has been almost that long since the miracle drugs (HMG CoA Reductase inhibitors or "statins") were made available for routine clinical use. If effective methods for diagnosing and treating FH have been available for over five years, why are very few patients getting properly diagnosed and treated?

One reason might be apathy toward the diagnosis and treatment of high cholesterol generated by debates about whether or not treating cholesterol really does any good. However it should be noted that even the most vocal critics of cholesterol testing and treatment agree with the need to diagnose and treat patients with FH.

Economic factors probably play a major role. Physicians largely do what they are paid to do. To effectively diagnose FH requires identifying and screening multiple relatives of an index case with high cholesterol. Most medical financing systems have not reimbursed physicians and their staff for the time-consuming job of tracing and screening relatives. Furthermore, many of these systems do not pay for the educational and support activities necessary to help

Table 3. Current status of diagnosis and treatment of FH

Country	Percent of FH patients diagnosed	Percent of FH patients on *any* medication	Percent of all FH patients with good[a] control of cholesterol
Australia	8	?	?
Austria	0.5	2	5
Canada	5	10	2
Denmark	10	10	3
Germany	10	15	< 5
Hungary	11	6	2
Israel	1.3	50	17
Netherlands	20	14	7
Norway	25	20	8
South Africa	< 10	< 3	< 1
Sweden	30	?	12
Switzerland	20	35	10
United Kingdom	10	10	5
United States	34	42	23
Average for MED-PED Countries	20	16	7

[a]"Good control" defined as treated cholesterol < 6.5 mmol or < 250 mg/dl.

FH patients and their physicians establish and maintain an effective treatment program.

Helping patients with FH requires the development of new approaches and tools not currently part of existing health care systems. MED-PED collaborators are addressing these obstacles to more effective diagnosis and treatment. If we cannot not achieve success with FH (a "best case scenario" single gene disease), then it would probably be even harder to find and help persons with other disorders.

MED-PED FH: A Practical Approach to Early Diagnosis, Treatment, and Prevention

The MED-PED approach includes several steps:

1) Collect a large sample of confirmed FH index cases and, with their permission, enter them into a computerized FH registry (in each country). These cases can be found using records from lipid clinics, laboratory surveillance, population cholesterol screening projects, and direct appeal to patients via news media.
2) Collect family questionnaires from these FH index cases to identify their close relatives who are also at high risk for FH.
3) Arrange to have someone contact the relatives to see if they have known high cholesterol levels or if they need to have their cholesterol level checked. The method of contacting relatives varies between countries. It may be done by the index case, by practicing physicians, or by paid or volunteer MED-PED staff.
4) Maintain a current computer registry (in each country) of consenting FH patients, their interested relatives and physicians (complete with addresses and phone numbers for follow up contacting).
5) Develop a registry of physicians with experience in treating FH.
6) Make regular contacts to FH patients in the registry to:
   a. give them educational materials to explain FH to them,
   b. encourage them to see a physician for treatment. (In some cases refer them to one of the registered FH physicians near them),
   c. contact them periodically (at least yearly) to see if they are taking medications and maintaining normal cholesterol levels.
7) Contact primary care physicians responsible for specific FH patients to:
   a. notify them that their patient has this serious disorder,
   b. offer to send free educational materials,
   c. provide consultative support services through the network of FH physicians through phone or personal consultations and encourage referrals and working collaboration between the primary care and FH physicians.

d. convince them that it is not acceptable to leave their FH patients untreated. They must either treat their patient effectively with MED-PED help, or refer their patient to a doctor with expertise in FH.

8) Work with lay organizations (like family heart associations), government offices, and media to increase public awareness of this treatable disorder and to develop stable funding for the activities needed to find and help patients with FH.

Ethical, Legal, and Social Implications

Genetic screening and contacting relatives require special attention to the protection of privacy and informed consent. In each case, the balance between the risks and benefits of screening must be considered (usually by some formal institutional review board for medical projects involving human participants). FH patients usually know about their strong positive family history for early heart attack deaths. Many of them even know about their high cholesterol levels. Most of them do not know their diagnosis is FH and most of them are not receiving competent medical help. Thus for persons with FH, the benefits of screening and family tracing far outweigh the risks. The methods for finding and contacting FH patients and relatives vary in different locations and reflect the ethical guidelines from local ethical review committees.

Once a person has the diagnosis of FH, he or she may be subjected to discriminatory practices regarding employment and insurance. MED-PED health professionals are working to help address these challenges in each country.

Initial Success with MED-PED

Table 4 lists the number of FH patients in local registries in the 14 countries collaborating in MED-PED. Several countries have been actively working with FH patients for several years before the MED-PED collaboration was formed (Hayden and Josephson 1993; Williams et al. 1993b). Others have just recently begun to collect FH patients in their registry. The 12,000 total FH patients currently in the combined MED-PED registries represent just 1% of the estimated 1.2 million FH patients in this group of countries. It is obvious that the work is just beginning. In the most experienced locations, a few index cases have routinely led to the identification of at least 2–5 other FH relatives. Analogous to searching for gold, each FH family is like a gold mine. It is more efficient to find more gold in existing mines that to search for new mines. In MED-PED the major emphasis is collecting *known* FH index cases and then contacting their relatives.

MED-PED FH a Paradigm for Other Dominant Single Gene Diseases

This same approach could work for similar diseases as long as they meet these criteria:

Table 4. Early progress of MED-PED FH registries

Country	Number of FH index cases	Number of relatives with FH	Total number of FH cases (Index + Rel)	Percent with known FH mutations
Australia	207	45	252	?
Austria	50	0	50	0
Canada	1,203	187	2,390	8
Denmark	100	200	300	85
Germany	420	1,080	1,500	10
Hungary	79	82	161	?
Israel	55	88	143	56
Netherlands	375	1,773	2,148	21
Norway	400	1,200	1,600	55
South Africa	1,150	127	1,277	30
Sweden	140	210	250	7
Switzerland	135	154	289	10
United Kingdom	100 +	100 +	200	30
United States	606	1,873	2,479	5
Total in MED-PED FH registries	5,103	7,315	13,318	?

1) a single dominant gene causes preventable serious illness.
2) validated diagnostic tests are available (gene test or clinical test).
3) some form of treatment or prevention is available and shown to be effective.

Several other single gene disorders that fit these criteria are presented in Table 5.

The Bottom Line: More Public Awareness and Financial Support are Needed

FH is about as common as AIDS. Tragic premature deaths occur often in both disorders. Moreover FH is currently more treatable than AIDS. While government agencies continue to fund important efforts for diseases like AIDS, they should also begin to invest in activities that will help prevent needless early deaths in persons with treatable genetic diseases like FH. Achieving this balanced government funding approach will require the coordinated efforts of a large number of FH patients and their physicians speaking in concert.

In the meantime, the initial efforts of MED-PED to help FH families are proceeding thanks to the combined support of several sources (Merck Human Health Inc. and its international affiliates, the World Health Organization, family and national heart associations in several countries, and some government agencies like state and provincial health departments and the US Centers for Disease Control).

Can you imagine the outrage that would arise if 600 jumbo jet airplanes crashed each year due to defective engine bolts that could have been repaired but were not? Then help us raise a similar outrage against the estimated 200,000

Table 5. Examples of treatable dominant "disease genes"

Genetic Trait	Description	Clinical diagnosis	Genetic diagnosis	Treatment or prevention
FH Familial hyper-choles-terolemia	Twice normal LDL Cholesterol and very early heart attack deaths (men in 20s to 50s, and woman about 10–20 years older)	High blood cholesterol in relatives and youth and or xanthoma. Quite reliable	a. Linkage to LDL receptor b. 160 causal mutations	Potent drugs can normalize cholesterol, and probably prolong life 10–30 years
FDB Familial defective Apo B	Twice normal apolipoprotien B and cholesterol with early heart attack deaths	High cholesterol levels can mimic FH, but some are lower	Two specific causal mutations	Same as for FH for those with very high cholesterol
Dominant Type III hyper-lipidemia	Very high beta VLDL cholesterol with early heart attack deaths	High triglyceride and abnormal Trig/VLDL ratio	Several specific causal mutations	Specific medications can often normalize levels and prolong life
Long QT syndrome	High risk for sudden arrhythmic death in youth and young adults	Long duration of QT interval on electro-cardiogram	Linkage found and 2 mutations found	Medication can lower risk of sudden death and prolong life
GRA hyper-tension	Severe high blood pressure and early stroke	Abnormal steroid hormones. BP normal after dexamethasone	Several specific causal mutations	Suppress abnormal steroids with hormones like dexamethasone
Breast cancer (BrCA 1)	Dominantly inherited breast cancer in families. Gene carriers also have increased risk of ovarian cancer	Mammography, clinical exams, and biopsy can diagnose cancer but not gene	36 causal mutations at one locus (BrCA 1), and linkage with second locus (BrCA 2)	Early detection and treatment or possibly prophylactic surgery can prevent fatal metastatic disease
Colon cancer	Dominantly inherited genes for colon cancer	Stool blood, X-ray tests, colonoscopy, and biopsy can diagnose cancer or polyps but not gene	Soon available	Early detection and treatment or possibly prophylactic surgery can prevent fatal metastatic disease

premature deaths worldwide in middle-aged men and women who die prematurely each year because their FH is not diagnosed or properly treated!

This effort must succeed so that we can make early diagnoses and prevent early deaths in medical pedigrees with FH!

References

Goldman L, Goldman PA, Williams LW, Weinstein ML (1993) Cost-effectiveness considerations in the treatment of heterozygous familial hypercholesterolemia with medications. Am J Cardiol 72: 75D–79D

Hayden MR, Josephson R (1993) Development of a program for identification of patients with familial hypercholesterolemia in British Columbia: A model for prevention of coronary disease. Am J Cardiol 72: 25D–29D

Hobbs HH, Brown MS, Goldstein JL (1992) Molecular genetics of the LDL receptor gene in familial hypercholesterolemia. Human Mutation 1: 445–446

Illingworth RD (1993) How effective is drug therapy in heterozygous familial hypercholesterolemia? Am J Cardiol 72: 54D–58D

Kane JP, Malloy MJ, Ports TA, Phillips NR, Diehl JC, Havel RJ (1990) Regression of coronary atherosclerosis during treatment of familial hypercholesterolemia with combined drug regimens. J Am Med Assoc 264: 3007–3012

Scandanavian Simvastatin Survival Study Group (1994) Randomised trial of cholesterol lowering in 4444 patients with coronary heart disease: the Scandinavian Simvastatin Survival Study (4S). Lancet 344: 1383–1389

Williams RR, Hunt SC, Schumacher MC, Hegale RA, Leppert MF, Ludwig EH, Hopkins PN (1993a) Diagnosing heterozygous familial hypercholesterolemia using new practical criteria validated by molecular genetics. Am J Cardiol 72: 171–176

Williams RR, Schumacher MC, Barlow GK, Hunt SC, Ware JL, Pratt M, Latham BD (1993b) Documented need for more effective diagnosis and treatment of familial hypercholesterolemia according to data from 502 heterozygotes in Utah. Am J Cardiol 72: 18D–24D

Yamamoto A, Kamiya T, Yamamura T, Yokoyama S, Horiguchi Y, Funahashi T, Kawaguchi A, Miyake Y, Beppu S, Ishikamu K et al. (1989) Clinical features of familial hypercholesterolemia. Arteriosclerosis 9 (Suppl I): I-66–I-74

Molecular Genetics of the Renin Angiotensin Aldosterone System in Human Hypertension

P. CORVOL, F. SOUBRIER, and X. JEUNEMAITRE

Blood pressure is a quantitative trait that varies continuously throughout the whole population and whose regulation is controlled by a variety of mechanisms that involve several genetic loci and environmental factors. However, little is known about the genes actually involved in human hypertension, about their respective importance in determining blood pressure level, and their interaction with other genes and environmental components. A number of epidemiologic studies have shown that individual blood pressure levels result from both genetic predisposition and environmental factors. The heritable component of blood pressure has been documented in familial and twin studies. The evidence suggests that approximately 30% of the variance of blood pressure is attributable to genetic heritability and 50% to environmental influences (Ward 1990).

With the notable exception of two autosomal dominant forms of hypertension related to either an abnormality in aldosterone secretion (glucocorticoid suppressible hyperaldosteronism; GSH) or an overactivity of the epithelial sodium channel (Liddle's syndrome), for which the molecular basis has been recently elucidated, there is no indication of the number of genetic loci involved in the regulation of blood pressure, the frequency of deleterious alleles, their mode of transmission, and the quantitative effect of any single allele on blood pressure. The unimodal distribution of blood pressure within each age group and in each sex strongly suggests, but does not definitively prove, that several loci are involved. Because of the likely etiologic heterogeneity of the disease, it is difficult to expect that a single biochemical or DNA genetic marker will help the clinician in the management of most hypertensive patients. However, genetic markers are useful indicators for elucidating the various genetic loci linked to high blood pressure. The genetic approach can disregard a gene as being frequently and importantly implicated in the level of blood pressure or in hypertension. The discovery of a positive linkage between a given locus and high blood pressure promotes new studies for finding a functional gene variant and for identifying new intermediate phenotypes of the locus. Finally, molecular genetics can unravel an underestimated or even totally unexpected mechanism of blood pressure control.

K. Berg, V. Boulyjenkov, Y. Christen (Eds.)
Genetic Approaches to Noncommunicable Diseases

Even though the genetic loci controlling blood pressure are unknown, a first and logical approach is to study genes that may contribute to the variance of blood pressure because of their well-known effect on the cardiovascular system. The genes of the renin angiotensin aldosterone system are a good illustration of such a "candidate gene" approach, since this system is well known to be involved in the control of blood pressure and in the pathogenesis of several forms of experimental and human hypertension. This system consists of four main proteins: renin, angiotensinogen, angiotensin II-converting enzyme (ACE) and angiotensin II receptors. During the past 10 years, considerable progress has been achieved, as all these genes have been cloned in humans and informative genetic markers have been identified. Recently, the genes coding for the enzymes involved in aldosterone biosynthesis and for the epithelial sodium channel have been cloned, allowing discovery of the gene mutations responsible for GSH and Liddle's syndrome, respectively.

This review will discuss recent progress made in the molecular basis of GSH and Liddle's syndrome, and in the molecular genetics of the renin angiotensin system genes in human hypertension.

Monogenic Forms of Hypertension Related to the Renin Angiotensin Aldosterone System

The molecular basis of two rare forms of severe hypertension have been recently elucidated. They are both characterized by an early onset of hypertension, frequent cardiovascular complications, an autosomal dominant inheritance, and, usually, a strong penetrance. In these two forms of hypertension, there is an extracellular volume expansion leading to a suppression of plasma renin activity, and a slight hypokaliemia with an inappropriately high kaliuresis due to the increased tubular Na^+/K^+ exchange in the distal tubule.

Glucocorticoid Suppressible Hypertension (GSH)

In GSH (also called dexamethasone-suppressible aldosteronism or glucocorticoid remediable aldosteronism), there is a variable degree of aldosteronism and an increased urinary excretion of 18-hydroxylated and oxocortisol metabolites, 18-hydroxycortisol and 18-oxocortisol. All these abnormalities, including hypertension, can be corrected by suppression of adrenocorticotropic hormone (ACTH) by dexamethasone. It had been proposed that the disease was due to an abnormal expression in the zona fasciculata of aldosterone synthase, CYP11B2, the enzyme responsible for the conversion of corticosterone into aldosterone. Recently, Lifton et al. (1992) showed that this disorder was linked to an abnormal aldosterone synthase gene. They studied a large kindred affected with this disease and found a gene duplication arising from unequal crossing over, resulting in a fusion of the 11β-hydroxylase (CYP11B1) promotor with the coding sequence of aldosterone synthase. The chimaeric gene was supposed to encode a protein that can hydroxylate cortisol (the steroid substrate present in the zona fasciculata) in the 18-position. This gene is under the control of the 11β-hydroxylase gene regulatory region, whose expression is under ACTH control and can be down-regulated by exogenous glucocorticoid administration. All the phenotypic abnormalities of GSH could be explained by this mutation.

Since this original description, several other families affected with this syndrome have been identified. In all families so far, the chimaeric gene is issued from unequal homologous recombination between exons 2 and 4 of the 11β-hydroxylase and aldosterone synthase genes, respectively. It has been shown by site-directed mutagenesis that such hybrid genes expressed in heterologous eucaryotic cells are able to encode chimeric proteins that hydroxylate cortisol in the 18-position (Pascoe et al. 1992). Finally, in a case of GSH with an adrenal tumor, we found expression of the chimeric gene in the tumoral tissue and in the zona fasciculata of the surrounding adrenal tissue (Pascoe et al., in preparation).

Liddle's syndrome

In 1963, Liddle et al. described a family with hypertension and an abnormality of Na^+ reabsorption at the level of the renal distal tubule simulating primary aldosteronism but with negligible basal and stimulated aldosterone secretion. Blood pressure and hypokaliemia were not influenced by spironolactone treatment, but triamterene, a specific inhibitor of the distal renal epithelial sodium channel, corrected these abnormalities. The authors proposed that the primary abnormality was a constitutive activation of the epithelial sodium channel. Some 30 years later, this hypothesis was revisited in the proband and in the originally described pedigree. The index case developed renal failure, and renal transplantation corrected the aldosterone and renin responses to salt restriction (which were blunted before), showing the involvement of the kidney in the disease (Botero-velez et al. 1994) and making the epithelial amiloride-sensitive sodium channel located in the collecting cortical tubule a logical candidate gene for this disease. This channel is constituted of at least three homologous subunits, α,β and γ, which act together to provide it its low sodium conductance and its high selectivity for sodium and amiloride. In a recent paper, Shimkets et al. (1994) showed complete linkage of the gene encoding the β subunit of this channel with the syndrome in the original pedigree. In this pedigree and in independent kindreds, a premature stop codon or a frameshift mutation was found in the intra-cellular carboxy-terminal domain of the β subunit. In addition, in a Portugese family affected with this syndrome, we found a 32 base pair deletion leading to a premature termination of the carboxy-end of the same subunit (Jeunemaitre and Corvol, unpublished). Interestingly enough, all these mutations lie in the same region of the β subunit and concern amino acids that may be phosphorylation sites or may interact with cytoskeleton proteins. They all induce a gain of function resulting in a constant activation of the channel.

New Questions from GSH and Liddle's Syndrome

Several interesting lessons as well new questions arise from these findings.

1) As shown by these two caricatural forms of hypertension, it is clear that single gene mutations can provoke hypertension. These conditions, as well as that of apparent mineralocorticoid excess (AME), which might be due to an abnormality of the 11β-HSD 2 gene, were considered to be rare causes of endocrine hypertension. However, it is likely that the prevalence of these diseases was underestimated due to the difficulty of establishing the diagnosis. In GSH and Liddle's syndrome, a single genetic test can be performed when the diagnosis is suspected which simplifies the diagnosis and may increase the actual number of cases diagnosed.
2) Although these two forms of hypertension are autosomal dominant with a strong penetrance, the effect of other genes acting on blood pressure and inherited from affected or non-affected parents can increase or eventually decrease the blood pressure level and modulate the phenotype among generations. It has been suggested that the impact of the GSH chimeric gene on blood pressure varies between families and even differs within a single family. Similarly, there is a large variability of the phenotypic expression of the disease in the original Liddle's pedigree, since some patients have a moderate hypertension and no hypokaliemia. Altogether, these findings suggest that the genetic variance of blood pressure level in these patients is the net result of "hypertensive genes" and "hypotensive genes" inherited from other members of the family.
3) Finally, these findings raise the possibility that similar genetic mechanisms operate is some common forms of essential hypertension with a related phenotype such as low plasma renin activity, variable hypokaliemia, and unappropriately high or low plasma aldosterone. The inheritable salt-dependent form of hypertension (Volpe et al. 1991) might represent the phenotype of another genetic abnormality of the epithelial sodium channel. This possibility should prompt studies looking at a possible association or linkage of these genes with high blood pressure in patients with a strong family history of hypertension and suggestive phenotypes.

Molecular Genetics of the Renin Angiotensin System in Human Essential Hypertension

The Renin Gene

The renin gene is important because the renin angiotensinogen reaction is the first and rate-limiting step leading to angiotensin II production. Numerous studies have involved renin to some degree in experimental forms of hypertension and in human hypertension. A fulminant hypertension develops in transgenic rates harboring the mouse Ren 2 gene (Mullins et al. 1990). Even more interesting was the pioneering observation by Rapp et al. (1989) of a cosegregation between a renin gene polymorphism and blood pressure level in a F2 population issued from crosses between inbred salt-sensitive and salt-resistant Dahl rats. In humans, about 30% of subjects with essential hypertension have higher renin levels than do normotensive subjects of the same age when examined under the same metabolic conditions (Brunner et al. 1972). Recently, it has been proposed that these patients are at higher risk of developing cardiovascular complications than normal-renin patients (Alderman et al. 1991); even though this remains a controversial issue (Meade et al. 1993).

The human renin gene is located on the short arm of chromosome 1 (1932–1942; Cohen-Haguenauer et al. 1983). Several restriction fragment length polymorphisms (RFLPs) have been located throughout the renin gene: TaqI and BglI polymorphisms in the 5′ region, Hind III in the 3′ region, and Hinf I in the first intron. In a single large Utah human pedigree with high prevalence of coronary disease and hypertension, there was no significant association between the renin RFLPs and blood pressure or plasma renin (Naftilan et al.). Interpretation was, however, limited by the very low number of patients studied. In another preliminary report, Morris and Griffiths (1988) compared the renin RFLPs of 29 subjects under antihypertensive treatment with those of 202 adult patients. No association was found between hypertension and the renin gene allele but, again, no definite conclusion could be drawn since only a few hypertensives were studied, clinical data were not available, and the renin gene polymorphism was defined by a single and weakly informative RFLP.

Soubrier et al. (1990) reported a study comparing the frequency of renin RFLPs in a large and contrasted population of normotensive and hypertensive subjects. A group of 102 hypertensive patients was selected according to strict criteria of age (between 20 and 60 years), established essential hypertension, and familial history of hypertension (defined as occurring before age 65 years in at least one parent and one sibling). A group of 120 normotensive subjects without a personal history of high blood pressure or a family history of hypertension was matched for age, sex ratio, and body-mass index. In all cases, patients presenting other hypertensive risk factors, such as elevated body-mass index, alcohol excess, oral contraceptive treatment, or diabetes mellitus, were excluded. Renin gene allele and haplotype frequencies were similar in the hypertensive and the normotensive groups.

To explore further the potential role of the renin gene as a genetic determinant of hypertension, Jeunemaitre and colleagues (1992a) used the hypertensive sib pairs approach. This methodology offers several advantages: 1) it does not assume any specific mode of inheritance at the test locus and looks only for a distortion of segregation between a genetic marker and the disease; 2) the use of a single adult generation decreases problems due to the age-related increase in blood pressure; and finally, 3) analysis of hypertensive sib pairs can partly resolve the problem of genetic heterogeneity generated by the analysis of extended multigenerational families. Using the same clinical criteria as Soubrier et al. (1990) and the same renin gene haplotypes, no linkage was found between the renin gene and hypertension, suggesting again that the renin gene does not have a frequent and/or important role in the pathogenesis of essential hypertension. However, the definitive exclusion of a contribution of the renin gene in the heritability of essential hypertension will require more powerful linkage studies, such as the use of a reliable renin intermediate phenotype, and a more polymorphic marker of the renin locus. It is not currently possible to exclude a minor role of this gene in blood pressure level in a large population of patients, or a major gene effect in rare families.

The Angiotensin I-converting Enzyme (ACE)

Angiotensin I-converting enzyme (ACE) is a zinc metalloprotease whose main functions are to convert angiotensin I into angiotensin II and to inactivate bradykinin. It is assumed that this step of the renin angiotensin system is not limiting in plasma, and indeed there is no indication that plasma ACE levels are directly related to blood pressure levels. However, the local generation of angiotensin I and the degradation of a bradykinin might depend on the level of ACE expressed in tissues.

Molecular cloning of the human endothelial ACE cDNA (18) has revealed that the enzyme consists of two highly homologous and functionally active domains resulting from a gene duplication. The organization of the human ACE gene provides further support for the duplication of an ancestral gene (Hubert et al. 1991). There are two ACE promoters, a somatic promoter localized on the 5′ side of the first exon of the gene and a germinal, intragenic promoter located on the 5′ side of the specific testicular ACE mRNA (Hubert et al. 1991; Kumar et al. 1991; Howard et al. 1990). The two alternate promoters of the ACE gene exhibit highly contrasting cell specificities, as the somatic promoter is active in endothelial, epithelial and neuronal cell types, whereas the germinal promoter is only active in a stage-specific manner in male germinal cells (Howard et al. 1990).

Relationship Between Plasma ACE Levels and Genotype

In a large series of normal individuals, Alhenc-Gelas et al. (1991) found plasma ACE levels differ markedly from subject to subject, from one to fivefold. ACE levels, however, remain remarkably constant when measured repeatedly in a given subject. This important variability is due, in large part, to a major genetic effect, as shown by Cambien et al. (1988) in a family population study where there was an intrafamilial resemblance between plasma ACE levels. This effect accounted for approximately 30% and 75% of the ACE variance in parents and in offsprings, respectively.

The role of the ACE gene in the genetic control of plasma ACE was assessed using ACE DNA polymorphism. A polymorphism consisting in the presence or the absence of a 287 base pair DNA fragment was detected and used as a marker genotype (24). In 80 healthy subjects, allele frequencies were 0.6 and 0.4 for the shorter (Deletion, D) and the longer allele (Insertion, I), respectively. Serum ACE levels were measured and classified according to ACE genotypes. Patients homozygous for the D allele had an immunoreactive ACE level almost twice as high as patients homozygous for the I allele, whereas heterozygous patients had an intermediate ACE level. This I/D polymorphism accounted for 47% of the total variance of serum ACE, showing that the ACE gene locus plays an important role in determining serum ACE levels (Rigat et al. 1990). As for serum ACE, T-lymphocyte ACE levels are significantly higher in patients homozygous for the D allele than in the other subjects (Costerousse et al. 1993). The ACE I/D polymorphism is not directly involved in the genetic regulation of serum and tissue ACE, and the causative variant responsible for the increase in ACE has yet to be found. Indeed, another study combining segregation and linkage analysis in 98 healthy nuclear families showed that the ACE I/D polymorphism is in fact only a neutral marker in strong linkage disequilibrium with the putative functional variant (Tiret et al. 1992). Altogether, these results suggest that the ACE expression level is under genetic control in cells synthesizing the enzyme, such as vascular or perivascular cells present in kidney, brain and heart.

ACE Gene Polymorphism and Hypertension

The observation that plasma ACE levels are under direct control of an ACE gene variant makes plausible the hypothesis that ACE is a possible candidate gene for high blood pressure. Two studies performed in genetically hypertensive rats rendered this hypothesis even more attractive. A F2 rat population generated from stroke-prone, spontaneously hypertensive rats (SHR/SP) and normotensive Wistar Kyoto crosses was studied by two laboratories, using a set of gene markers evenly spaced throughout the rat genome (Hilbert et al. 1991; Jacob et al. 1991). Both groups of investigators found a significant linkage between NaCl-loaded hypertension and a gene locus on rat chromosome 10 that contributed as much as 20% of blood pressure variance under high salt intake. The ACE gene was contained within the large confidence interval of the locus. Interestingly, the human ACE gene maps in a syntenic region of the rat genome, at band 17q23 (Mattei et al. 1989).

One association study comparing a normotensive and a hypertensive Australian population with two hypertensive parents, showed an association of hypertension with ACE gene polymorphism (Zee et al. 1992). In fact, the significant difference between the I/D genotype originated only from a subgroup of patients aged 50 years or more, where the D genotype was less frequent than in normotensives. This finding was interpreted as due to an over-risk of cardiovascular events in hypertensive patients carrying the D allele (Morris et al. 1994). Harrap et al. (1993) investigated the distribution of the ACE I/D gene polymorphism described above in young adults with contrasting genetic predisposition to high blood pressure (Watt et al. 1992; "four-corners approach"). Young adults with high blood pressure and two parents with high blood pressure did not show any significant difference in the I/D allele frequencies of the ACE gene when compared with adults that were the same age but had low blood pressure, and no genetic predisposition to high blood pressure. Other association studies were also negative (Higashimori et al. 1993; Schmidt et al. 1993). The most extensive study was that of Jeunemaitre et al. (1992b); who showed no evidence of linkage between hypertension and a growth hormone gene polymorphic marker in complete linkage disequilibrium with the ACE gene in hypertensive sib pairs from Utah. Taken together, these results suggest that the ACE gene does not play a major role on blood pressure variance in these populations.

ACE Gene Polymorphism and Other Diseases

It is interesting to note that the ACE I/D polymorphism seems to be a potent risk factor for coronary heart disease, especially in patients formerly considered at low risk according to common criteria. A case control study (ECTIM) was performed in different populations in France and in Ireland to identify variants of candidate genes predisposing to myocardial infarction. Cambien et al. (1992) found that, when compared to ID and II genotypes, the ACE DD genotype was associated with an excess of cases with myocardial infarction in a low-risk group defined according to normal plasma ApoB levels and body mass index. In this subgroup analysis, the ACE I/D polymorphism was found to be an independent risk factor: the ACE DD genotype increased 2.7 times the relative risk (approximated by odds ratio) of developing a myocardial infarction. In a subsequent study, the same investigators showed that there was a significant excess of both DD and ID genotypes among patients with a parental history of fatal myocardial infarction (Tiret et al. 1993). In a preliminary report, Otishi et al. (1993) showed an increased incidence of coronary artery restenosis following coronary artery dilation in patients carrying the ACE DD genotype when compared to the II genotype. Other studies are presently in progress in several countries to test the hypothesis that the ACE I/D gene polymorphism could be a new and potent risk factor for coronary diseases.

Two recent and still preliminary studies suggest that the ACE DD genotype frequency is increased in patients with ischaemic or idiopathic dilated cardiomyopathy (Raynolds et al. 1993) and in familial hypertrophic cardiomyopathy (Marian et al. 1993); favoring the hypothesis that an ACE gene variant may contribute to the pathogenesis of these diseases and that a genegene interaction might play a detrimental role in familial hypertrophic cardiomyopathy and sudden cardiac death. Finally, a population-based study performed in 141 women and 149 with left ventricular hypertrophy indicates that the DD genotype of ACE is a genetic marker of left ventricular hypertrophy, especially in middle-aged men (Schunkert et al. 1994).

Recent data also suggest that the ACE genotype may be associated with diabetes mellitus complications. Insulin-dependent diabetic patients with diabetic nephropathy exhibited the II genotype significantly less frequently than their control subjects in two independent studies (Marre et al. 1994; Doira et al. 1994). Similarly, the D allele of the ACE gene was found to be a strong and independent risk factor for coronary heart disease in non-insulin-dependent diabetic patients (Ruiz et al. 1994). In this study, which included 316 patients of which 132 had coronary heart disease, there was a progressively increasing relative risk in heterozygous and homozygous for the D allele, suggesting a codominant effect on cardiovascular complications.

The association of cardiovascular risk with the DD genotype in different clinical situations leads to the hypothesis that the deleterious effect of the DD genotype results from an overexpression of ACE producing a local increase in angiotensin II in some vascular territories like the coronary and the renal circulation. However, several words of caution should be expressed in the interpretation of these findings: 1) although the DD genotype is clearly associated with an increase in plasma and T lymphocyte ACE levels in normotensives, there is so far no indication that a parallel increase exists in hypertensives (Higashimore et al. 1993) and in tissues such as the heart, the kidney or the vascular wall; 2) all studies reported so far are association studies with serious limitations, in terms of number of patients, selection bias of patients and controls, and limited statistical power; 3) the results of association analysis should also be interpreted in the context of geographical and racial background, since a difference has been shown in the ACE I/D genotypes according to the ethnic origin of populations (Barley et al. 1994; Lee 1994; Ishigami et al. 1995; Duru et al. 1994); and 4) the functional variant in strong linkage disequilibrium with the ACE I/D polymorphism has not yet been found. It may lie within the ACE gene itself but we cannot exclude the possibility that it is another gene, physically close to the ACE gene.

The Angiotensinogen Gene

Angiotensinogen, the renin substrate, is mainly synthesized by the liver and is the unique substrate for renin. Although the concentration of angiotensinogen present in plasma seems relatively high, it is actually within a range where variations of its concentration directly affects the angiotensin I production rate. Because the latter is one-half maximal at the usual plasma angiotensinogen concentration, it is logical to suspect that a chronic state of increased plasma angiotensinogen might facilitate hypertension and/or cardiovascular diseases. Its role in human hypertension was suspected in an epidemiological study, where a strong correlation was found between plasma angiotensinogen concentration and blood pressure (Walker et al. 1979); and in another study where offspring of hypertensive patients had elevated plasma angiotensinogen levels (Fasola et al. 1968).

The human angiotensinogen gene belongs to the superfamily of serpins (Gaillard et al. 1989) and is localized to chromosome 1q42.3, in the same region as human renin (Gaillard-Sanchez et al. 1990).

Essential Hypertension

An extensive study of the potential role of the angiotensinogen gene in human essential hypertension was performed in two large series of hypertensive sibships yielding a total of 379 sib pairs (Salt Lake City, Utah, USA and Paris, France; Jeunemaitre et al. 1992c). Using a highly polymorphic GT microsatellite located in the 3′ region of the angiotensinogen gene (Kostelevstev et al. 1991); evidence of genetic linkage between the angiotensinogen gene and hypertension was obtained. An excess of angiotensinogen allele sharing was found in severely hypertensive sibpairs (characterized by a diastolic blood pressure greater than 100 mmHg or taking two or more antihypertensive medications). In both the Utah and Paris groups, although a significant linkage was obtained in male pairs, no excess of shared angiotensinogen alleles was observed in female comparisons, suggesting the influence of an epistatic hormonal phenomenon.

The genomic sequences of the angiotensinogen locus were analyzed further by multiple PCR fragments and mutations searched in coding and non-coding regions. Among the 15 observed variants, five were missense mutations and three were observed in single families. Two frequent variants, 174M and 235T, were found to be both linked to and associated with hypertension. The M235T variant (Met--> Thr in amino-acid position 235) was found more frequently in hypertensive probands, especially in the more severe index cases (0.50),than in controls (0.38) in both Caucasian groups. Finally, a significant increase in plasma angiotensinogen level was observed for patients bearing the M235T variant, with a 10 and 20% increase in heterozygotes (MT) and homozygotes (TT), respectively, compared to wild-type homozygotes MM (Jeunemaitre et al. 1992c). More recently, these results were confirmed in hypertensive patients without selection for a strong family history of hypertension (Jeunemaitre et al. 1993). The corroboration and replication afforded by these results support the interpretation that molecular variants of angiotensinogen, such as M235T or tagged by this variant, constitute inherited predispositions to essential hypertension in humans.

Two recent studies support a relationship between the angiotensinogen locus and high blood pressure. In the United Kingdom, Caulfield et al. (1994) showed a strong linkage and an association of the angiotensinogen gene locus to essential hypertension in a set of British families, despite the fact there was no association between hypertension and the M235T variant, which could be due to population differences. The effect of the angiotensinogen gene polymorphism on blood pressure was studied in a genetic isolate, the Hutterite Brethren, in North America. A group of 741 Hutterites was tested for association between systolic and diastolic blood pressure and between M235T and T174M genotypes. There was a significant association in resting systolic blood pressure and 174M polymorphism, only in men. The angiotensinogen 174M polymorphism accounted for 3% of the variance in systolic blood pressure in men (Hegile et al. 1994). The 235T allele frequency varies between populations and reaches up to 70% in the African American population and 81% in Nigerians. There was no relation between 235T or 174M allele frequencies and hypertension in a study of African Americans (Rotoimi et al. 1994). However, in the Japanese population, where the 235T allele frequency is also higher than in the Caucasian population, a significant association between this allele and high blood pressure was found (Hata et al. 1994; Kamitani et al. 1994; Iwai et al. 1994).

Pregnancy-Induced Hypertension

The previous studies support the hypothesis of a susceptible allele of the angiotensinogen gene that is associated with an increase of both plasma angiotensinogen concentration and blood pressure, an effect that could be more striking in conditions of stimulation of angiotensinogen expression, such as oral estrogen administration or pregnancy. Clinical studies have documented a familial tendency to develop preeclampsia, and familial studies have suggested both a genetic inheritance and the influence of environmental factors (Cooper et al. 1988; Liston and Kilpatrick 1991).

Two recent reports indicate that the angiotensinogen locus could play an important role in the occurrence of pregnancy-induced hypertension. Ward et al. (1993) found a significant association between the angiotensinogen 235T variant and preeclampsia in both Caucasian and Japanese samples. Using another strategy, analysis of the allelic inheritance of the GT repeat in 52 sibling pairs of preeclamptic sisters, Arngrimsson et al. (1993) showed a significant linkage between the angiotensinogen locus and preeclampsia in Icelandic and Scottish families. Thus, although different mechanisms have been proposed in preeclampsia and essential hypertension, these results suggest that some common variants of the angiotensinogen gene could predipose to both diseases.

From all these studies, the angiotensinogen gene appears to be involved in the determinism of human familial hypertension and some forms of pregnancy-induced hypertension. However, as for all predisposing genes for a common disease, other clinical studies will have to be conducted in different populations and races to ascertain its role in high blood pressure. Finally, several questions will have to be resolved: 1) the effect of the angiotensinogen locus on blood pressure is unknown but might be weak and modulated by a variety of interacting loci and environmental factors, accounting for the negative results in some under powered studies; 2) it is not possible to determine at the present time whether the observed molecular variants of angiotensinogen directly affect angiotensinogen function or whether they are markers of functional variants which have not yet been detected; 3) if, indeed, the mutation of a methionine into a threonine in the 235 position directly affects plasma angiotensinogen concentration, it will be necessary to look for a possible difference in clearance rate or Km for renin between the two angiotensinogen isoforms; and 4) the response to antihypertensive agents, especially those blocking the renin system, will have to be evaluated in patients classified according to their angiotensinogen genotype.

The Angiotensin II Type 1 Receptor

Angiotensin II receptors, which mediate all the biological and physiological effects of the renin angiotensin system, are also candidate genes for essential hypertension. The AT_1 subtype is a G-coupled receptor located to the plasma membrane of angiotensin II target cells: vascular smooth muscle cells, renal vasculature and mesangial cells, adrenal and brain. The human gene has been cloned and is located on chromosome 3 (Furuta et al. 1992; Curnow et al. 1992).

In a first series of experiments, putative molecular variants in the coding region of the gene that might be functional were sought. Such variants in other G-coupled seven-transmembrane domain receptors have been reported for the human thyrotropin (Parma et al. 1993) and luteinizing hormone (Shenker et al. 1993) receptors; resulting in a constitutive activation or hyperresponsiveness of the receptor. In the case of hypertension similar functional variants of the angiotensin II receptors could lead to some forms of essential hypertension or to tumoral primary aldosteronism, in the case of sporadic tissular mutations. No evidence of such mutations in the coding region of the AT_1 receptor gene was found in 60 probands of hypertensive families (Bonnardeaux et al. 1994) and in 20 cases of tumoral primary aldosteronism (Davies et al., in preparation). However, an informative diallelic marker $A^{1166} \rightarrow C$ was found in the 3′ untranslated region of the AT_1 gene. A significant increase in the allelic frequency of this variant was more frequently present in hypertensive subjects, suggesting that a variant of AT_1 receptor exerts a small effect on blood pressure (Bonnardeaux et al. 1994). Interestingly, this gene polymorphism may interact synergistically with the ACE D allele in the ECTIM study, mentioned earlier, for increasing the relative risk for myocardial infarction (Tiret et al. 1994).

Conclusion

Molecular genetic studies make possible an evaluation of the contributions of the renin angiotensin system genes to blood pressure variance and hypertension in animals and in humans. It is currently likely that neither the renin nor the ACE genes contribute greatly to genetic hypertension, at least in humans. However, a subset of the human population that has yet to be defined, could still be involved. Molecular variants of angiotensinogen gene constitute inherited predispositions to essential hypertension in humans and are probably involved in some cases of pregnancy-induced hypertension. Finally, an ACE gene polymorphism associated with an increase in plasma and tissular ACE levels appears to be a strong marker of coronary and cardiac disease.

All these results show that some molecular variants of angiotensinogen (235T) or of ACE (DD genotype) are associated with an increased plasma (and may be tissular) angiotensinogen and ACE levels, respectively. This could result, in turn, in a small increase in the formation rate of angiotensin II, especially in tissues where these proteins are rate limiting for angiotensin II generation. This genetically chronic overstimulation of the renin system would then favor kidney sodium reabsorption, vascular hypertrophy and (or) increase sympathetic nervous system activity and predispose to the development of common cardiovascular diseases. This effect might be more marked in the presence of other predisposing genes and/or deleterious environmental factors.

Acknowledgments. This work was supported by Grants from INSERM, Collège de France, Bristol-Myers Squibb, Association Claude Bernard and Association Naturalia and Biologia.

References

Alderman MH, Madhavan S, Ooi WL, Cohen H, Sealey JE, Laragh JH (1991) Association of the renin-sodium profile with the risk of myocardial infarction in patients with hypertension. New Engl J Med 324: 1098–1104

Alhenc-Gelas F, Richard J, Courbon D, Warnet JM, Corvol P (1991) Distribution of plasma angiotensin I converting enzyme levels in healthy men: relationship to environmental and hormonal parameters. J Lab Clin Med 117: 33–39

Arngrimsson R, Purandare S, Connor M, Walker JJ, Björnsson S, Soubrier F, Kotelevtsev Y, Geirsson RT, Björnsson H (1993) Angiotensinogen: a candidate gene involved in preeclampsia. Nature Genet 4: 114–115

Barley J, Blackwood A, Carter ND, Crews DE, Cruickshank JK, Jeffery S, Ogunlesi AO, Saguella GA (1994) Angiotensin converting enzyme insertion/deletion polymorphism: association with ethnic origin. J Hypertens 12: 955–957

Bonnardeaux A, Davies E, Jeunemaitre X, Fery I, Charru A, Tiret L, Cambien F, Corvol P, Soubrier F (1994) Angiotensin II type 1 receptor gene polymorphism in human essential hypertension. Hypertension 24: 63–69

Botero-velez M, Curtis JJ, Warnock DG (1994) Liddle's syndrome revisited–A disorder of sodium reabsorption in the distal tubule. New Engl J Med 330: 178–181

Brunner HR, Laragh JH, Baer L (1972) Essential hypertension: renin and aldosterone heart attack and stroke. New Engl J Med 286: 441–449

Cambien F, Alhenc-Gelas F, Herbeth B, André JL, Rakotovao R, Gonzales MF, Allegrini J, Bloch C (1988) Familial resemblance of plasma angiotensin converting enzyme levels. Am J Human Genet 43: 774–780

Cambien F, Poirier O, Lecert L, Evans A, Cambou JP, Arveller D, Luc G, Bard JM, Bara L, Ricard S, Tiret L, Amouyel P, Alhenc-Gelas F, Soubrier F (1992) Deletion polymorphism in the gene for angiotensin-converting enzyme is a potent risk factor for myocardial infarction. Nature 359: 641–644

Caulfield M, Lavender P, Farrall M, Munroe P, Lawson M, Turner P, Clark AJL (1994) Linkage of the angiotensinogen gene to essential hypertension. New Engl J Med 330: 1629–1633

Cohen-Haguenauer O, Soubrier F, N'Guyene VC, Serero S, Turleau C, Jegou C, Gross MS, Corvol P, Frezal J (1983) Regional mapping of the human renin gene to $1q^{32}$ by in situ hybridization. Ann Genet (Paris) 32: 16–20

Cooper DW, Hill JA, Chesley LC (1988) Genetic control of susceptibility to eclampsia and miscarriage. Br J Obstet Gynecol 95: 644
Costerousse O, Allegrini J, Lopez M, Alhenc-Gelas F (1993) Angiotensin I-converting enzyme in human circulating mononuclear cells. Genetic polymorphism of expression in T-lymphocytes. Biochem J 290: 33–40
Curnow K, Pascoe L, White PC (1992) Genetic analysis of the human type-1 angiotensin II receptor. Mol Endocrinol 6: 1113–1118
Doira A, Warram JH, Krolewski AS (1994) Genetic predisposition to diabetic nephropathy. Evidence for a role of the angiotensin I-converting enzyme gene. Diabetes 43: 690–694
Duru K, Farrow S, Wang JM, Lockette W, Kurtz T (1994) Frequency of a deletion polymorphism in the gene for angiotensin converting enzyme is increased in African-Americans with hypertension. Am J Hypertens 7: 759–762
Fasola AF, Martz BL, Helmer OM (1968) Plasma renin activity during supine exercise in offsprings of hypertensive parents. J Appl Physiol 25: 410–415
Furuta H, Deng-Fu G, Inagami T (1992) Molecular cloning and sequencing of the gene encoding human angiotensin II type 1 receptor. Biochem Biophys Res Commun 183: 8–13
Gaillard I, Clauser E, Corvol P (1989) Structure of human angiotensinogen gene DNA 8: 87–89
Gaillard-Sanchez I, Mattei MG, Clauser E, Corvol P (1990) Assignment by in situ hybridization of angiotensinogen to chromosome band 1q42: the same region as human renin gene. Human Genet 84: 341–343
Harrap SB, Davidson RH, Connor MJ, Soubrier F, Corvol P, Fraser R, Foy CJW, Watt GCM (1993) The angiotensin I converting enzyme gene polymorphism, predisposition to high blood pressure and activity of the renin-angiotensin system. Hypertension 21: 455–460
Hata A, Namikawa C, Saski M, Sato K, Nakamura T, Tamura K (1994) Angiotensinogen as a risk factor for essential hypertension in Japan. J Clin Invest 93: 1285–1287
Hegile RA, Brunt H, Connelly PW (1994) A polymorphism of the angiotensinogen gene associated with variation in blood pressure in a genetic isolate. Circulation 90: 2207–2212
Higashimori K, Zhao Y, Higaki J, Kamitani A, Katsuya A, Nakuyra J, Miki T, Mikami H, Ogihara T (1993) Association of a polymorphism of the angiotensin converting enzyme gene with essential hypertension in the Japanese population. Biochem Biophys Res Commun 191: 393–404
Hilbert P, Lindpaintner K, Beckmann JS, Serikawa T, Soubrier F, Dubay C, Cartwright P, De Gouyon B, Julier C, Takahasi S, Vincent M, Ganten D, Georges M, Lathrop GM (1991) Chromosomal mapping of two genetic loci associated with blood pressure regulation in hereditary hypertensive rats. Nature 353: 521–526
Howard TE, Shai SY, Langford KG, Martin BM, Bernstein KE (1990) Transcription of testicular angiotensin-converting enzyme (ACE) is initiated within the 12th intron of the somatic ACE gene. Mol Cell Biol 10: 4294–4302
Hubert C, Houot AM, Corvol P, Soubrier F (1991) Structure of the angiotensin I-converting enzyme gene: two alternate promoters correspond to evolutionary steps of a duplicated gene. J Biol Chem 266: 15377–15383
Ishigami T, Iwamoto T, Tamura K, Yagamuchi S, Iwasawa K, Uchino K, Umemura S, Ishii M (1995) Angiotensin I converting enzyme (ACE) gene polymorphism and essential hypertension in Japan. Ethnic difference of ACE genotype. Am J Hypertens 8: 95–97
Iwai N, Ohmichi N, Nakamura Y, Mitsunnami X, Kinoshita M (1994) Molecular variants of the angiotensinogen gene and hypertension in a Japanese population. Hypertens Res 17: 117–121
Jacob HJ, Lindpaintner K, Lincoln SE, Kusumi K, Bunker RK, Mao YP, Ganten D, Dzau VJ, Lander ES (1991) Genetic mapping of a gene causing hypertension in the stroke prone spontaneously hypertensive rat. Cell 67: 213–224

Jeunemaitre X, Rigat B, Charru A, Houot AM, Soubrier F, Corvol P (1992a) Sib pair linkage analysis of renin gene haplotypes in human essential hypertension. Human Genet 88: 301–306
Jeunemaitre X, Lifton RP, Hunt SC, Williams RR, Lalouel JM (1992b) Absence of linkage between the angiotensin-converting enzyme locus and human essential hypertension. Nature Genet 1: 72–75
Jeunemaitre X, Soubrier F, Kotelevtsev Y, Lifton RP, Williams CS, Charru A, Hunt SC, Hopkins PN, Williams RR, Lalouel J-M, Corvol P (1992c) Molecular basis of human hypertension. Role of angiotensinogen. Cell 71: 169–180
Jeunemaitre X, Charru A, Chatellier G, Dumont C, Sassano P, Soubrier F, Mènard J, Corvol P (1993) M235T variant of the human angiotensinogen gene in unselected hypertensive patients. J Hypertens 11 (Suppl 5): S80–S81
Kamitani A, Rakugi H, Higaki J, Yi Z, Miki T, Ogihara T (1994) Association analysis of a polymorphism of the angiotensinogen gene with essential hypertension in Japanese. J Human Hypertens 8: 521–524
Kotelevstev YV, Clauser E, Corvol P, Soubrier F (1991) Dinucleotides repeat polymorphism in the human angiotensinogene gene. Nucl Acids Res 19: 6978
Kumar RS, Thekkumkara TJ, Sen GC (1991) The mRNAs encoding the two angiotensin-converting isozymes are transcribed from the same gene by a tissue-specific choice of alternative transcription initiation sites. J Biol Chem 266: 3854–3862
Lee EJD (1994) Population genetics of the angiotensin-converting enzyme in Chinese. Br J Clin Pharmacol 37: 212–214
Liddle GW, Bledsoe T, Coppage WS (1963) A familial renal disorder simulating primary aldosteronism but with negligible aldosterone secretion. Trans Assoc Am Physicians 76: 199–213
Lifton RP, Dluhy RG, Powers M, Rich GM, Cook S, Ulick S, Lalouel JM (1992) A chimaeric 11ß-hydroxylase/aldosterone synthase gene causes glucocorticoid-remediable aldosteronism and human hypertension. Nature 355: 262–265
Liston WA, Kilpatrick DC (1991) Is genetic susceptibility to pre-eclampsia conferred by homozygosity for the same single recessive gene in mother and fetus? J Obstet Gynecol 98: 1079–1086
Marian AJ, Yu QT, Workman R, Greeve G, Roberts R (1993) Angiotensin-converting enzyme polymorphism in hypertrophic cardiomyopathy and sudden cardiac death. Lancet 342: 1085–1086
Marre M, Bernadet P, Gallois Y, Savagner F, Guyene TT, Hallab M, Cambien F, Passa P, Alhenc-Gelas F (1994) Relationship between angiotensin I converting enzyme gene polymorphism, plasma levels, and diabetic retinal and renal complications. Diabetes 43: 384–388
Mattei MG, Hubert C, Alhenc-Gelas F, Rolckel N, Corvol P, Soubrier F (1989) Angiotensin I converting enzyme is on chromosome 17. 10th International Workshop on Human Gene Mapping, New Haven, Connecticut, June 11–17, 1989. Cytogenet Cell Genet 51: 1041
Meade TW, Cooper JA, Peart WS (1993) Plasma renin activity and ischemic heart disease. New Engl J Med 329: 619
Morris BJ, Griffiths LR (1988) Frequency in hypertensives of alleles for a RELP associated with the renin gene. Biochem Biophys Res Commun 150: 219–224
Morris BJ, Zee RYL, Schrader AP (1994) Different frequencies of angiotensin-converting enzyme genotypes in older hypertensive individuals. J Clin Invest 94: 1085–1089
Mullins JJ, Peters J, Ganten D (1990) Fulminant hypertension in rats harboring the mouse Ren 2 gene. Nature 344: 541–544
Naftilan AJ, Williams R, Burt D, Paul M, Pratt RE, Hobart P, Chirgwin J, Dzau VJ (1989) A lack of linkage of renin gene restriction length polymorphisms with human hypertension. Hypertension 14: 614–618

Otishi M, Fujii K, Minaminot T, Higaki J, Kamitani A, Rakugi H, Zhao Y, Hikami H, Miki T, Ogihara T (1993) A potent genetic risk factor for restenosis. Nature Genet 5: 324–325

Parma J, Duprez L, Van Sande J, Cochaux P, Gervy C, Mockel J, Dumon J, Vassart G (1993) Somatic mutations in the thyrotropin receptor gene cause hyperfunctioning thyroid adenomas. Nature 365: 649–651

Pascoe L, Curnow KM, Slutsker L, Connell JMC, Speiser PW, New MI, White P (1992) Glucocorticoid-suppressible hyperaldosteronism results from hybrid genes created by unequal crossovers between CYP11B1 and CYP11B2. Proc Natl Acad Sci USA 89: 8327–8331

Rapp JP, Wang SH, Dene H (1989) A genetic polymorphism in the renin gene of Dahl rats cosegregates with blood pressure. Science 243: 542–544

Raynolds MC, Bristow MR, Bush EW, Abraham WT, Lowes BD, Zisman LS, Taft CS, Perryman MB (1993) Angiotensin-converting enzyme DD genotype in patients with ischaemic or dilated cardiomyopathy. Lancet 342: 1073–1075

Rigat B, Hubert C, Alhenc-Gelas F, Cambien F, Corvol P, Soubrier F (1990) An insertion/deletion polymorphism in the angiotensin I converting enzyme gene accounting for half the variance of serum enzyme levels. J Clin Invest 86: 1343–1346

Rotoimi C, Morrison L, Cooper R, Oyejide C, Effiong E, Lapids M, Osotemihen B, Ward R (1994) Angiotensinogen gene in human hypertension: Lack of an association of 235T allele among African Americans. Hypertension 24: 591–594

Ruiz J, Blanché H, Cohen N, Velho G, Cambien F, Cohen D, Passa P, Froguel P (1994) Insertion/deletion polymorphism of the angiotensin-converting enzyme gene is strongly associated with coronary heart disease in non-insulin-dependent diabetes mellitus. Proc Natl Acad Sci USA 91: 3662–3665

Schmidt S, Van Hoof IMS, Grobbee DE, Ganten D, Ritz E (1993) Polymorphism of the angiotensin I converting enzyme gene is apparently not related to high blood pressure: Dutch hypertension and offspring study. J Hypertension 11: 345–348

Schunkert H, Hense H-W, Holmer SR, Stender M, Perz S, Keil U, Lorell BH, Riegger GA (1994) Association between a deletion polymorphism of the angiotensin-converting enzyme gene and left ventricular hypertrophy. New Engl J Med 330: 1634–1638

Shenker A, Laue L, Kasugi S, Merendino JJ, Minegishi T, Cutler GB (1993) A constitutively active mutation of the luteinizing hormone receptor in familial male precocious puberty. Nature 365: 652–654

Shimkets RA, Warnock DG, Bositis CM, Nelson-Williams C, Hansson JH, Schambelan M, Gill JR, Ulick S, Milora RV, Findling JW, Canessa C, Rossier BC, Lifton RP (1994) Liddle's syndrome: Heritable human hypertension caused by mutations in the β subunit of the epithelial sodium channel. Cell 79: 407–414

Soubrier F, Alhenc-Gelas F, Hubert C, Allegrini J, John M, Tregar G, Corvol P (1988) Two putative active centers in human angiotensin I converting enzyme revealed by molecular cloning. Proc Natl Acad Sci (USA) 85: 9386–9390

Soubrier F, Jeunemaitre X, Rigat B, Houot AM, Cambien F, Corvol P (1990) Similar frequencies of renin gene restriction fragment length polymorphisms in hypertensive and normotensive subjects. Hypertension 16: 712–717

Tiret L, Rigat B, Visvikis S, Breda C, Corvol P, Cambien F, Soubrier F (1992) Evidence from combined segregation and linkage analysis that a variant of the angiotensin I-converting enzyme (ACE) gene controls plasma ACE levels. Am J Human Genet 51: 197–210

Tiret L, Kee F, Poirier O, Micaud V, Lecerf L, Evans A, Cambon JP, Arveiler D, Luc G, Amoyel P, Cambien F (1993) Deletion polymorphism in angiotensin-converting enzyme gene associated with parental history of myocardial infarction. Lancet 341: 991–993

Tiret L, Bonnardeaux A, Poirier O, Ricard S, Marques-Vidal P, Evans A, Arveiler D, Luc G, Kee F, Ducimetierre P, Soubrier F, Cambien F (1994) Synergistic effects of angiotensin converting enzyme and angiotensin II type 1 receptor gene polymorphisms on risk of myocardial infarction. Lancet 344: 910–913

Volpe M, Lembo G, De Luca N, Lamenza F, Tritto C, Ricciaderli B, Molaro M, De Campera P, Condorelli G, Rendina V, Trimarco B, Condorelli M (1991) Abnormal hormonal and renal responses to saline load in hypertensive patients with parental history of cardiovascular accidents. Circulation 84: 92–100

Walker WG, Whelton PK, Saito H, Russel R, Hermann J (1979) Relation between blood pressure and renin, renin substrate, angiotensin II, aldosterone and urinary sodium and potassium in 574 ambulatory subjects. Hypertension 1: 287–291

Ward R (1990) Familial aggregation and genetic epidemiology of blood pressure. In: Laragh JH, Brenner BM (eds), Hypertension: pathophysiology, diagnosis and management. Raven Press, New York, pp 81–100

Ward K, Hata PF, Jeunemaitre X, Helin C, Nelson L, Namikawa C, Farrington PF, Ogasawara M, Suzumori K, Tomoda S, Berrebi S, Sasaki M, Corvol P, Lifton R, Lalouel JM (1993) A molecular variant of angiotensinogen associated with preeclampsia. Nature Genet 4: 59–61

Watt GCM, Harrap SB, Foy CJW, Holton DW, Edwards HV, Davidson R, Connor JM, Lever AF, Fraser R (1992) Abnormalities of glucocorticoid metabolism and the renin angiotensin system: a four-corners approach to the identification of genetic determinants of blood pressure. J Hypertension 10: 473–482

Zee RYL, Lou YK, Griffiths LR, Morris BJ (1992) Association of an insertion/deletion polymorphism of the angiotensin I-converting enzyme gene with essential hypertension. Biochem Biophys Res Commun 184: 9–15

Genetics of Non-insulin-dependent Diabetes Mellitus Among Mexican Americans: Approaches and Perspectives

C.L. HANIS

Abstract

Pedigree investigations, population studies, animal models and molecular genetic studies consistently implicate a substantial role of genes in determining susceptibility to non-insulin-dependent diabetes mellitus (NIDDM). They also point to a likely genetic model involving several genes. As with other common chronic conditions, the number of genes involved, their locations, and the causative mutations have thus far proven elusive. Identification of these loci is complicated by the fact that mutations at these loci are neither necessary or sufficient to produce disease. Molecular biology and technology now permit a complete search of the genome to resolve these issues. Once accomplished, it will be possible to determine the interactions of these genes with environmental factors in determining susceptibility to NIDDM and its complications in individuals, families and populations.

Introduction

It is no longer a rare occurrence to see reported the elucidation of the underlying genetic mechanisms of a variety of diseases; indeed, hardly a month goes by without one or more such reports. These reports include the mapping of disease genes, their cloning and identification of causative mutations. Examples include cystic fibrosis (Rommens et al. 1989), Huntington's disease (Huntington's Disease Collaborative Research Group 1993), and muscular dystrophy (Koenig et al. 1987). Efforts with these diseases are now turning toward exploitation of this information for early identification and the development of treatment strategies. Common to those conditions for which success is most often reported is the simple nature of their inheritance and the relative infrequency in most populations. Similar successes have not been reported for such common chronic conditions as cardiovascular disease, hypertension, diabetes and obesity. Will the same advances in molecular biology and genetic mapping that have led to

K. Berg, V. Boulyjenkov, Y. Christen (Eds.)
Genetic Approaches to Noncommunicable Diseases

the former successes lead to similar findings for these latter conditions? The answer appears to be both yes and no. Yes, in the sense that the genes will be found and this will lead to an understanding of underlying biological mechanisms and the identification of individuals with increased disease susceptibility. The answer will also be no in that it is probable that several genes will be found, with no one being either necessary or sufficient to cause disease. This complexity may preclude the direct application of the paradigm that moves from linkage to positional cloning and subsequent treatment. However, the sheer impact of the common chronic diseases on populations calls out for the development of similar strategies. In this report, we discuss the problems and opportunities associated with identifying disease-susceptibility loci for non-insulin-dependent diabetes mellitus (NIDDM) among Hispanic Americans in Starr County, Texas.

Starr County, Texas, is one of a handful of US counties that border with Mexico. It is located on the Rio Grande River approximately 240 km upstream from where it empties into the Gulf of Mexico. For at least the past two decades, Starr County has had the highest diabetes-specific mortality of any county in Texas. In the state of Texas, approximately 18 diabetes deaths are reported for every 1000 deaths. In Starr County, this number is 52 (see Hanis et al. 1983). Not only does Starr County have high diabetes mortality, but so do most counties in Texas that border with Mexico. In fact, counties with a high proportion of Hispanics (predominantly Mexican Americans) have increased diabetes-specific mortality. The population of Starr County is 98% Mexican American. Subsequent surveys of the population have established that NIDDM is three- to five-fold more frequent in Starr County than in the general US population (Hanis et al. 1983). Similar findings have been shown for other Mexican American populations in Texas (Stern et al. 1981, 1984), New Mexico (Samet et al. 1988), Colorado (Hamman et al. 1989), and in general (Flegal et al. 1991).

Genetic Susceptibility to NIDDM

Evidence indicates that the increased frequency of NIDDM among Mexican Americans is due to an increased underlying genetic susceptibility interacting with diet, culture, lifestyle, etc. The Native American populations of North, Central and South America have also been shown to have elevated frequencies of NIDDM, even higher than those observed among Mexican Americans. For example, the Pima Indians of Arizona have among the highest frequencies of NIDDM of any population (Knowler et al. 1978). That these findings are consistent across widely varying environmental strata is strong evidence for underlying genetic susceptibility loci (Neel 1982; Weiss et al. 1984). If true, then any contemporary population whose gene pool is in part derived from Native American ancestry would be expected to have frequencies of NIDDM in proportion to the degree of gene sharing with Native Americans. This is precisely the pattern that is seen. It is estimated that 31% of the contemporary Mexican

American gene pool is Native American-derived, 8% from African ancestry and the remaining from Spanish ancestry (Hanis et al. 1991). Using just these admixture estimates and the frequencies of diabetes in the "ancestral" populations, one can predict the expected frequency of NIDDM in the Mexican American population. This is illustrated in Fig. 1, where there is an unmistakable parallelism of the observed and expected frequencies. That the observed rates are systematically below those predicted may reflect other systematic differences in environmental factors.

Not only do population studies clearly implicate genetic susceptibility to NIDDM, but so do family studies. NIDDM clearly aggregates in families. Concordance rates for identical twins are more than 90% for NIDDM (Foster 1989). In Starr County we have collected data on more than 1,500 individuals distributed in some 150 pedigrees ascertained through a proband with NIDDM. Risks in male and female siblings of probands were 3.1 and 3.3 times higher than those expected for random males and females, respectively. Similarly, relative risks of NIDDM in male and female offspring of probands were 3.4 and 2.9, respectively. These results are comparable to those reported from studies of the Pima Indians. There, the risk to an offspring of one diabetic parent was elevated 2.3 times, whereas NIDDM risk was 3.9-fold higher in offspring of two NIDDM parents (Knowler et al. 1981). Relative risks for spouses of NIDDM probands in Starr County were 2.0 and 1.8 for males and females, respectively. These data clearly show that aggregation of NIDDM in pedigress is strongest for those who share both genes and common environmental factors with probands (siblings and offspring) compared to those who share only environmental commonalities (spouses). That there is substantial risk in this population for spouses of probands underscores the role of environmental factors and leads to

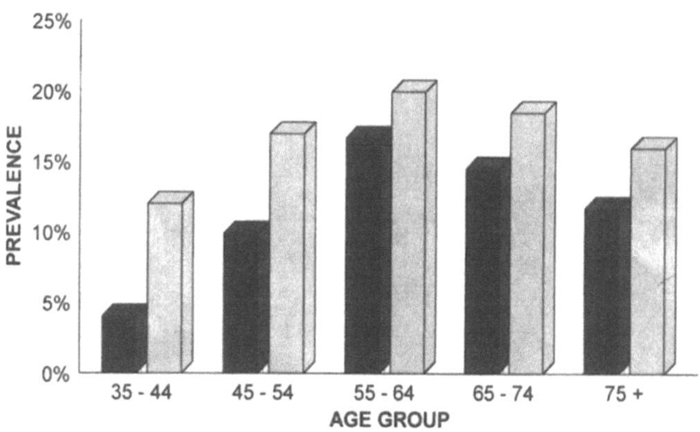

Fig. 1. Parallelism of observed NIDDM frequencies in Starr Country (*dark bars*) compared to frequencies predicted by genetic admixture alone (*light bars*). (Patterned after Fig. 3 of Hanis et al. 1991)

the supposition of complex interactions of environmental and genetic components.

Before leaving the issue of familial aggregation, it is important to note that not only does NIDDM aggregate in families but so do the complications of diabetes. Figure 2 documents the familial aggregation of diabetic retinopathy that we have observed in Starr County. The left panel shows the risk for NIDDM in siblings of probands based on whether or not the proband has retinopathy. Individuals with retinopathy may have a more severe from of NIDDM, and this may have a stronger underlying genetic basis. As seen, there are no differences in the risk of NIDDM as a function of retinopathy status of the proband. The right panel documents the risk of retinopathy in diabetic siblings of probands according to proband retinopathy status. There is more than a 9-fold increased risk of retinopathy in siblings of probands who have retinopathy compared to siblings of those who do not. Understanding of the sources of aggregation of complications is compounded by the requirement that NIDDM must first be present and NIDDM itself has a complex aggregation pattern.

Analysis of the Starr County pedigrees fails to resolve the mode of inheritance of NIDDM. Rather, no simple genetic model adequately explains the observed pattern of NIDDM aggregation in pedigrees. The default complexity then implies that several genes are likely to be involved. This is the general conclusion of several recent reviews on the genetics of NIDDM (Hamman 1992; Permutt 1990; Rich 1990).

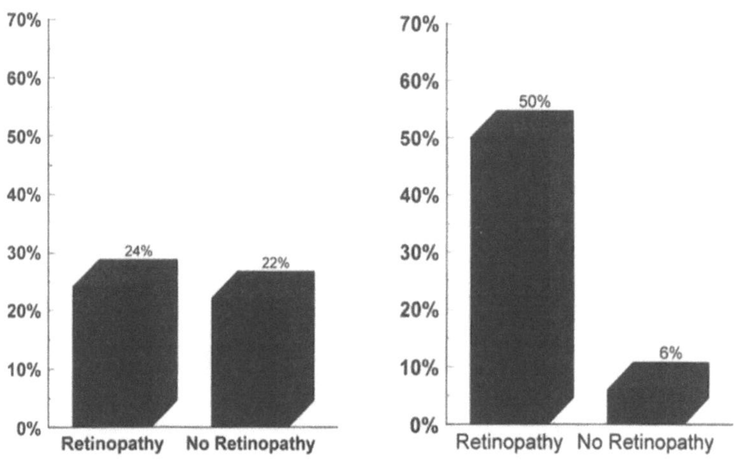

Fig. 2. Familial aggregation of diabetic retinopathy among Mexican Americans with NIDDM in Starr County, Texas. The *left panel* shows the risk for NIDDM in siblings of probands based on whether or not the proband has retinopathy. The *right panel* documents the risk of retinopathy in diabetic siblings of probands according to proband retinopathy status

Searching for NIDDM Genes

The nature of the genetic complexity underlying NIDDM has important implications for locating putative genes. Given that multiple loci are probably involved, there are three basic models that could explain the genetic contributions of NIDDM. First is a model in which there is one locus having a relatively major effect while the remaining loci have small effects. This appears to be the best explanation for insulin-dependent diabetes mellitus (IDDM) genetic susceptibility. The HLA locus has the single largest impact on IDDM risk. Next is the insulin gene region, but its effect is nearly an order of magnitude less than that of the HLA region. Several other regions have also been shown to be linked, but the impact in each case appears to be minor (Davies et al. 1994; Owerbach and Gabbay 1995). The second general model type would be the classical polygenic model in which there are many loci with small effects. Last is a model in which there are a few loci having small to moderate effects. It is generally presumed that the latter model is most appropriate for NIDDM (Rich 1990). However, the number of such genes, their locations and the mechanisms by which they increase susceptibility are not known. Neither is it known how these genes may interact with each other and with environmental factors. Finally, such a model implies that the loci involved will not be individually *necessary or sufficient* to produce diabetes. There will be heterogeneity from individual to individual and from pedigree to pedigree in the constellation of susceptibility loci that are segregating. It is imperative, though, that these genes be found in order to understand the events that lead to NIDDM and its complications. The recent advances in molecular biology and the detailed map of the human genome now available indicate that it is a propitious time to undertake searching the genome for NIDDM genes.

Two basic strategies are available for searching the genome for NIDDM loci. These are illustrated in Fig. 3, where the different approaches are indicated along with a bar representing the relative amount of genotyping required for each approach. The first is a "candidate" approach. Classically, this approach consists of detecting genetic variation at loci that are known to be involved in the metabolic events that are most involved with disease. Obvious candidates are the insulin gene, insulin receptor, and glucose transporters. Once variation is found, linkage and/or association studies are performed to determine what, if any, role the locus has in NIDDM. To date, dozens of candidate genes have been examined. Genetic variation has invariably been found and tests for linkage and association to NIDDM have been performed in a variety of populations. In no case has there been clear and unmistakable identification of a candidate gene as being an "NIDDM gene" (see Hamman 1992 for a review). This is still a viable and the most focused approach. The number of loci to be analyzed is simply the number of candidate genes postulated. This method has proven successful in IDDM with regard to the HLA locus and has also been successful in identifying the glucokinase gene on chromosome 7 in some pedigrees segregating for maturity onset diabetes of the young (MODY; Vionnet et al. 1992; Stoffel et al. 1992).

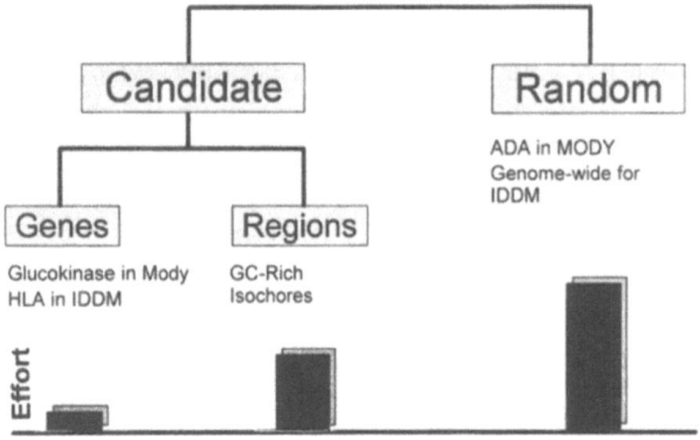

Fig. 3. Strategies for searching the human genome for NIDDM susceptibility loci

A variant of the candidate gene approach is a candidate region approach. Candidate regions are defined as areas of the genome holding clusters of genes or, in the broadest sense, areas that are likely to hold genes. These latter are identified as CpG islands and are enriched for holding genes (Craig and Bickmore 1994). This approach requires several-fold more locus-tests, but does not require the same intensity as a complete genome search. To date, there are no clear applications of this approach. This may be due to the rapidity with which the genetic map has developed and the ability to automate portions of the genotyping efforts. The proportionate savings in effort of this approach over a complete genome search are not nearly as large as those obtained by examining only candidate genes. At some point, this regional approach would also require examination of all other areas, which then involves a complete genome search using random markers.

The random marker approach is the more labor-intensive of the three strategies, but has the advantage that, in principal, all genes can be found. It is the approach generally used for mapping rare Mendelian diseases and was used to identify linkage of other MODY pedigrees to chromosome 20 near the adenosine deaminase locus (Bell et al. 1991). More recently, this approach has been used to identify IDDM susceptibility loci (Davies et al. 1994). In that study, 289 random markers were typed. Of these, 20 showed statistical significance: HLA and the insulin gene regions and 18 others. By testing these markers in different pedigree samples, they were able to confirm two possible other IDDM susceptibility loci.

Sibling Pair Linkage Approaches

The next five years will witness complete genome searches for NIDDM genes as well as other common chronic disease genes. The likelihood that these genome searches will identify the NIDDM susceptibility loci depends on several factors, including an appropriate sampling design, statistical procedures chosen, and

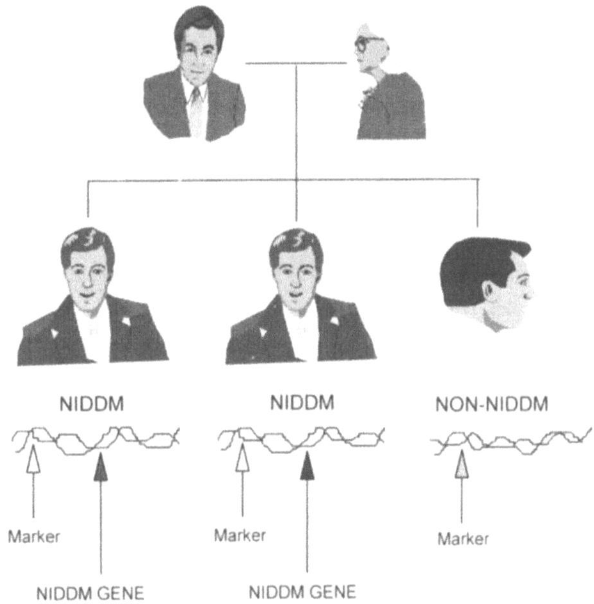

Fig. 4. A pedigree representation of the affected sibling pair method of linkage analysis

genetic marker density. Widespread interest and attention is being given to affected relative pair designs, especially affected sibling pairs. Affected sib pairs seem ideally suited as a design. They are relatively easy to collect, enrich for genetic causation and their analysis is straightforward. Statistical analysis of the affected pair data relies on the proportion of alleles shared identical by descent (IBD).Figure 4 represents a simple pedigree. Two offspring have NIDDM and the third does not. Assuming that genes are important for NIDDM, it is reasonable to suppose that the two siblings with NIDDM have inherited the same form of a putative NIDDM gene. Not only will they be similar for the NIDDM locus, but they will also have similar sequences at all loci near the NIDDM locus. Thus, there should be an increased sharing of alleles IBD at linked marker loci. As the distance from the disease locus increases, recombination will lead to an IBD distribution for the siblings that is the same for non-linked loci (i.e., 0, 1 or 2 alleles IBD with probabilities of 1/4, 1/2, and 1/4, respectively). This basic idea was originally formulated by Penrose (1953). The major advantage of affected pair methods is that they require no assumptions regarding the nature of the genetic model underlying susceptibility. This is critical for the common diseases because adequate genetic models have not been formulated. This precludes the application of more traditional parametric linkage approaches that are sensitive to model specification yet still provides adequate power to detect linkage (Risch 1990a, b).

Statistical testing based on identity by descent distributions has been refined. The most common implementation is the Affected Pair Method (APM) of Weeks and Lange (1988). In this method, siblings are not only scored for the number of alleles that are shared IBD, but the IBD distribution is weighted by a function of allele frequencies. Doing so gives greater weight to the sharing of rare alleles. Ideally, parents of the affected sibs are also genotyped to unequivocally establish IBD. This is not as applicable for a disease like diabetes, with a later age of onset. Collecting only pedigrees in which the parents are available is not only much more difficult but may lead to subtle biases in the type of diabetes that is seen, with undue weighting to those pedigrees with early onset diabetes. These may not be representative of the general population of individuals with NIDDM.

An alternative to IBD approaches are methods based on identity by state (IBS) distributions. Here, sib pairs are simply scored as to the number of alleles for which they are similar. Lange (1986) showed that little information is lost substituting IBS for IBD when sufficiently polymorphic markers are employed. This poses no problems today because of the abundance of short sequence repeat polymorphisms with multiple alleles and high heterozygosities (above 70%). Bishop and Williamson (1990) have developed a direct implementation of the IBS method based on the observed and expected distribution of 0, 1 or 2 alleles shared identical by state, employing a chi-square test of significance.

Probability of Success in Searching the Genome

Fundamental to the success of any genome searches is sufficient statistical power to detect linkage when, in fact, it exists. Estimates from the literature suggest that 200 affected sib pairs will provide more than 80% power under a variety of genetic models (Risch 1990b). In particular Risch has shown that 80% power will be obtained from 200 sib pairs when a disease producing locus increases risk in relatives by a factor of 2 and the marker locus is no more than 5 to 10 recombination units (cM) away. Based on the familial aggregation data reviewed, this may be at the upper limit of the size of effects to be found for NIDDM. As the distance between the marker and disease locus increases, power decreases dramatically.

To determine the power for detecting linkage under more likely situations, we have simulated data under a three-locus additive threshold model. It is assumed that each locus has two alleles and that one allele at each locus results in increased disease risk. Four risk-raising alleles are required for manifestation of disease. The frequencies of the risk-raising alleles at each locus were set to 0.42 so that the overall frequency of disease would be 21%. This approximates the lifetime risk of NIDDM of Mexican Americans in Starr County. A marker locus with six equally frequent alleles was simulated with various recombination distances from one of the disease loci. The polymorphic information content (PIC) value for this hypothetical marker is 0.81 and is well within the range of markers available. From a design point of view, heterogeneity can be diminished through application of a narrow, consistent disease diagnosis, but by definition, this additive model is a heterogeneity model in that the impact of each locus will differ among sib pairs and from pedigree to pedigree. It also has the characteristics that alleles at any one locus are not necessary or sufficient. The chi-square test of Bishop and Williamson was used to determine significance of results. This simulation assumed a sample of 300 affected sibling pairs. The simulation results are presented in Fig. 5, where the power of the statistical test as a function of θ is given (solid lines). Each point plotted represents the proportion of significant tests obtained in more than 3,000 replications. When $\theta = 0$, linkage will be detected 83% of the time. At 5% recombination, the power to detect linkage is approximately 70%. Doubling the sample size (data not shown), generates power above 80% even at 10% recombination.

To have a marker locus no more than 10% recombination units away from the disease locus requires testing markers spaced 20 cM from each other. It is estimated that the human genome spans some 4,000 cM (Morton 1991). A total of 200 markers would need to be typed to achieve the density above. The number of markers required goes up drastically as the maximal distance to a disease locus decreases. This is illustrated by the dashed line in Fig. 5. As the distance becomes smaller, the increased power achieved from doubling the marker density is minimal. In practice, it appears that some 400 to 600 markers evenly spanning the genome will provide sufficient power to detect disease loci having small to moderate effects.

ConÆrming Putative Linkages

Obtaining statistical significance from a linkage test as part of a complete genome search does not imply that a disease locus has been found. The result must be balanced against the likelihood of achieving such test by chance alone. In a search using 400 genetic markers, one can nominally expect 20 statistically

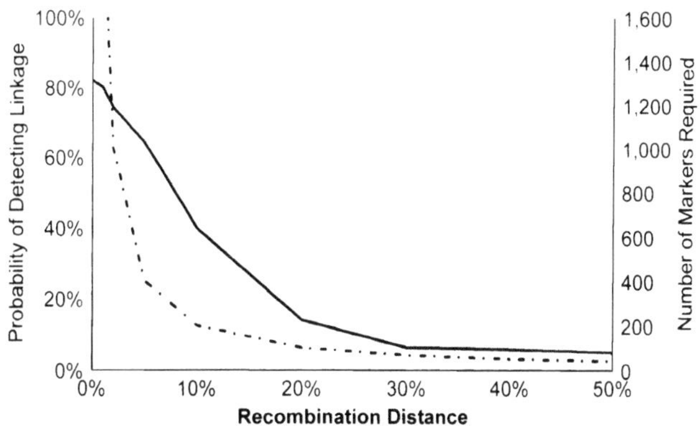

Fig. 5. Power to detect linkage using 300 affected sibling pairs (*solid line*) and the number of markers needing testing (*dashed line*) as a function of recombination distance (θ)

significant results by chance alone (employing a 5% critical value). It becomes essential, then, to separate those tests that are significant by chance alone and those that represent real linkages. A hierarchical strategy for confirming linkage is presented in Fig. 6. Minimally, the results need to be biologically consistent. In the case of sib pair approaches, linkage should lead to increased sharing of alleles by siblings. It is difficult to conceive of any mechanisms whereby linkage to a putative locus would lead to decreased allele sharing by siblings. Even if there are protective alleles, there will still be an increased sharing of the alternative alleles in affected individuals. Next, the results should be statistically consistent. If a disease locus is present, then there should be a strengthening of statistical significance as one gets closer to the disease locus, followed by diminished significance as the disease locus is passed. Note, however, that if a region is shared among siblings more often than by chance, then so too will any regions close to it. Therefore, these first two criteria are not considered strong evidence for confirming linkage.

A much more persuasive argument for linkage can be obtained by confirming the linkage in other data sets. Ideally, one would want to confirm the linkage in an independent sample from the same ethnic group and then in a different ethnic group. Not demonstrating linkage in the second ethnic group may reflect genetic heterogeneity among populations rather than a true absence of linkage. This is the strategy that Davies et al. (1994) used in confirming linkages for IDDM. At the same level in the hierarchy, one could also test for linkage in pedigrees. Having a putative linkage allows more flexibility in testing linkage in pedigrees using more traditional LOD score based methods. The ultimate confirmation of linkage is the identification of a causative mutation. Once that is achieved the opportunities for understanding are greatly increased.

Conclusions

The identification of NIDDM genes is not an end in and of itself. The ultimate purposes of locating these genes is to obtain fundamental understanding of the biological processes that lead to increased diabetes susceptibility, identify high

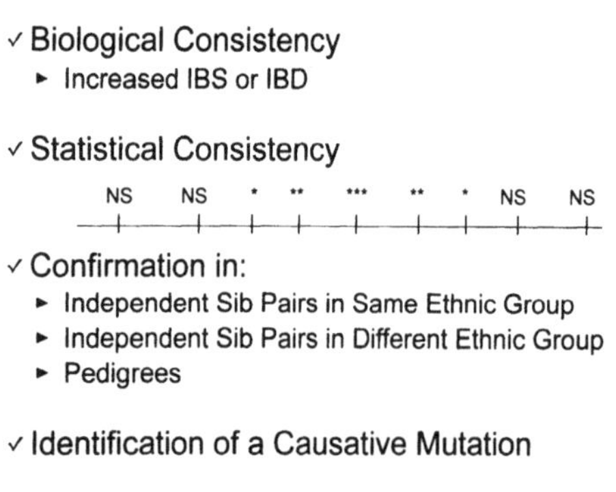

Fig. 6. A hierarchical strategy for confirming putative linkages detected in a complete genome search

risk individuals, families and populations, and develop intervention strategies to prevent or slow the process whereby health gives way to disease. Each of these endeavors first requires identification of the genes involved and the risk-raising alleles. Once these are discovered, a variety of lines of research will then be pursued. In the population context, two endeavors merit special attention: the first is examining the relationship of these loci to both disease and complications and the second is determining the interactions of these loci with environmental factors.

Three types of loci may be discovered. The first type consists of loci that increase diabetes susceptibility and also increase risk for developing the complications of diabetes. The second is loci that may increase diabetes susceptibility, but have no effect on the development of complications. Lastly are loci that may have direct effects on complication risk, but only once diabetes itself has been manifested. The last two groups may contribute to the variation in risk for diabetes complications among individuals, with some individuals being genetically susceptible or resistant to specific complications. If genes exist that only relate to complications, then an alternative design will need to be employed that incorporates concordance among sibling pairs for the complications of diabetes.

Although it has been implied throughout, it has not been specifically stated that there must be interaction of genetic susceptibility loci with environmental factors. It is thought that native Americans and those genetically related to them have higher frequencies of diabetes susceptibility alleles that are only manifest in a "diabetes-prone" environment (Weiss et al. 1984). Using the Pima Indians as an example, the frequency of diabetes has increased dramatically in this century. The forces that change gene frequencies do not produce substantial changes in the compressed period of only a few generations. Environmental changes in lifestyle, physical activity and diet have changed in this population and have produced the diabetes burden today. That there must first be an underlying genetic susceptibility for this increase in diabetes is established by observations of other populations that have undergone "westernization" of lifestyle and have had only minor increases in diabetes risk (see Weiss et al. (1984) for a review). For native Americans this implies that today's birth cohort could have diabetes risks that are equivalent to those of a century ago, given a less diabetogenic environmental exposure. Identification of the genes for diabetes will permit us to determine those environmental exposures that are diabetogenic through studies of genotype and environmental interactions. These interactions may be amenable to public health interventions in a fashion that will lower the ultimate impact of diabetes to levels not seen for decades.

Before we can resolve these critical issues, we must have the genes in hand. They will be found in the next few years. Several groups are now engaged in complete genome searches. Utilizing data on 346 Mexican American NIDDM-affected sib pairs from Starr County, we are systematically searching the entire genome at approximately a 5 cM level. To do so, we have entered into a consortium of four universites and have divided the genome into approximately equal components. Graeme Bell (University of Chicago, Chicago, Illinois) is scanning chromososmes 1 through 4, Patrick Concannon (University of Washington, Seattle, Washington) is scanning chromosomes 5 through 11, we are searching chromosomes 12 through 17, and Richard Spielman (University of Pennsylvania, Philadelphia, Pennsylvania) is searching the remaining chromosomes. This effort is approximately 80% complete and will identify NIDDM loci. The understanding that will follow will lead to identification of individuals at increased risk for developing NIDDM and its complications, to the development of new pharmaceutical approaches to diabetes management, and intervention strategies that exploit the interaction of genes and environmental factors. These developments will have strong, positive impacts on the substantial portion of the population sharing in the burden of non-insulin-dependent diabetes.

References

Bell GI, Xiang KS, Newman MB, Wu SH, Wright LG, Fajans SS, Spielman RS, Cox NJ (1991) Gene for non-insulin-dependent diabetes mellitus (maturity-onset diabetes of the young subtype) is linked to DNA polymorphism on human chromosome 20q. Proc Natl Acad Sci USA 88: 1484–1488

Bishop DT, Willamson JA (1990) The power of identity-by-state methods for linkage analysis. Am J Human Genet 46: 254–265

Craig JM, Bickmore WA (1994) The distribution of CpG islands in mammalian chromosomes. Nature Genetics 7: 376–382

Davies JL, Kawaguchi Y, Bennett ST, Copeman JB, Cordell HJ, Pritchard LE, Reed PW, Gough SCL, Jenkins SC, Palmer SM, Balfour KM, Rowe BR, Farrall M, Barnett AH, Bain SC, Todd JA (1994) A genome-wide search for human type 1 diabetes susceptibility genes. Nature 371: 130–136

Flegal KM, Ezzati TM, Harris MI, Haynes SG, Juarez RZ, Knowler WC, Perez-Stable EJ, Stern MP (1991) Prevalence of diabetes in Mexican Americans, Cubans, and Puerto Ricans from the Hispanic Health and Nutrition Examination Survey, 1982–1984. Diabetes Care 14 (suppl.3):628–638

Foster DW (1989) Diabetes Mellitus. In: Scriver C, Beaudet AL, Sly WS, Valle D (eds) The metabolic basis of inherited disease. New York, McGraw Hill, 375–397

Hamman RF (1992) Genetic and environmental determinants of non-insulin-dependent diabetes mellitus (NIDDM). Diabetes/Metabolism Rev 8: 287–338

Hamman RF, Marshall JA, Baxter J, Kahn LB, Mayer EJ, Orleans M, Murphy JR, Stamler J (1989) Methods and prevalence of non-insulin-dependent diabetes mellitus in a biethnic Colorado population: the San Luis Valley Diabetes Study. Am J Epidemiol 124: 295–311

Hanis CL, Ferrell RE, Barton SA, Aguilar L, Garza-Ibarra A, Tulloch BR, Garcia CA, Schull WJ (1983) Diabetes among Mexican-Americans in Starr County, Texas. Am J Epidemiol 118: 659–672

Hanis CL, Hewett-Emmett D, Bertin TK, Schull WJ (1991) The origins of U.S. Hispanics: Implications for diabetes. Diabetes Care 14 (suppl.3): 618–627

Huntington's Disease Collaborative Research Group (1993) A novel gene containing a trinucleotide repeat that is expanded and unstable on Huntington's disease chromosomes. Cell 72: 971–983

Koenig M, Hoffman EP, Berelson CJ, Monaco AP, Feener C, Kunkel LM (1987) Complete cloning of the Duchenne muscular dystrophy (DMD) cDNA and preliminary genomic organization of the DMD gene in normal and affected individuals. Cell 50: 509–517

Knowler WC, Bennett PH, Hamman RF, Miller M (1978) Diabetes incidence and prevalence in Pima Indians: 19-fold greater incidence than in Rochester, Minnesota. Am J Epidemiol 108: 497–505

Knowler WC, Pettitt DJ, Savage PJ, Bennett PH (1981) Diabetes incidence in Pima Indians: contribution of obesity and parental diabetes. Am J Epidemiol 113: 144–156

Lange K (1986) The affected sib-pair method using identity by state relations. Am J Human Genet 39: 148–150

Morton NE (1991) Parameters of the human genome. Proc Natl Acad Sci, USA 88: 7474–7476

Neel JV (1982) The thrifty genotype revisited. In: Köbberling J, Tattersall R (eds) The genetics of diabetes mellitus. New York, Academic Press, pp 283–293

Owerbach D, Gabbay KH (1995) The HOXD8 locus (2q31) is linked to type I diabetes: Interaction with chromosome 6 and 11 diseases susceptibility genes. Diabetes 44: 132–136

Penrose LS (1953) The general purpose sib-pair linkage test. Ann Eugen 18: 120–124

Permutt MA (1990) Genetics of NIDDM. Diabetes Care 13 (suppl.4): 1150–1153

Rich SS (1990) Mapping genes in diabetes: genetic epidemiological perspective. Diabetes 39: 1315–1319

Risch N (1990a) Linkage strategies for genetically complex traits. I. Multilocus models. Am J Human Genet 46: 222–228

Risch N (1990b) Linkage strategies for genetically complex traits. II. The power of affected relative pairs. Am J Human Genet 46: 229–241

Rommens JM, Lannuzzi MC, Kerem B, Drumm ML, Melmer G, Dean M, Rozmahel R, Cole JL, Kennedy D, Hidaka N, Zsiga M, Buchwald M, Riordan JR, Tsui LC, Collins FS (1989) Identification of the cystic fibrosis gene: Chromosome walking and jumping. Science 245: 1059–1065

Samet JM, Coultas DB, Howard CA, Skipper BJ, Hanis CL (1988) Diabetes, gallbladder disease, obesity and hypertension among Hispanics in New Mexico. Am J Epidemiol 128: 1302–1311

Stern MP, Gaskill SP, Allen CR, Garza V, Gonzales JL, Waldrop RH (1981) Cardiovascular risk factors in Mexican Americans in Laredo, Texas. I. Prevalence of overweight and diabetes and distributions of serum lipids. Am J Epidemiol 113: 546–555

Stern MP, Rosenthal M, Haffner SM, Hazuda HP, Franco LJ (1984) Sex difference in the effects of sociocultural status on diabetes and cardiovascular risk factors in Mexican Americans–the San Antonio Heart Study. Am J Epidemiol 120: 834–851

Stoffel M, Patel P, Lo YM, Hattersley AT, Lucassen AM, Page R, Bell JI, Bell GI, Turner RC, Wainscoat JS (1992) Missense glucokinase mutation in maturity-onset diabetes of the young and mutation screening in late-onset diabetes. Nature Genet 2: 153–156

Vionnet N, Stoffel M, Takeda J, Yasuda K, Bell GI, Zouali H, Lesage S, Velho G, Iris F, Passa P, Froguel P, Cohen D (1992) Nonsense mutation in the glucokinase gene causes early-onset non-insulin-dependent diabetes mellitus. Nature 356: 721–722

Weeks DE, Lange K (1988) The affected-pedigree-method of linkage analysis. Am J Human Genet 42: 315–326

Weiss KW, Ferrell RW, Hanis CL (1984) A new world syndrome of metabolic diseases with a genetic and evolutionary basis. Yearbook Phys Anthropol 27: 153–178

The Genetics of Asthma

W. Cookson

Introduction

Asthma

Asthma is the most common disease of childhood. One in seven children in the United Kingdom suffers from the illness (Strachan et al. 1994). Adult asthma is a significant cause of morbidity, costing the National Health Service about £400,000,000 per annum. Lost production causes a similar cost to the community. There are approximately 2000 asthma deaths per year in Great Britain, and this number has remained stable despite improvements in diagnosis and treatment. Similar figures are found throughout the Western world.

Asthma is a disease of the small airways of the lung. Intermittent narrowing of the respiratory bronchioles produces airflow limitation and the symptoms of wheeze and shortness of breath. Asthma is almost certainly not one disease but many. The most common form of asthma is allergic asthma, also known as atopic asthma. The term atopy, meaning "strange disease", was invented by Coca and Cooke in 1926 to describe a familial syndrome of asthma, seasonal rhinitis (hay fever) and infantile eczema. Ninety-five percent of childhood asthma is atopic. The prevalence of childhood asthma rises to a peak in the early teens, and is more common and earlier in onset in boys than in girls. The clinical course in many children is that of gradual improvement as they enter adulthood. Many individuals, unfortunately, remain disabled throughout their lives.

In adults most cases of asthma are also atopic, but many individuals with adult-onset asthma have no evidence of underlying atopy. In these cases, cigarette smoking is often, but not invariably, a contributing cause. Other asthma syndromes, with known precipitants, are also recognised. Aspirin-sensitive asthma affects 10% of adult onset asthmatics, and although aspirin ingestion leads to asthma in these patients, it does not seem to be through classical allergic pathways. Industrial asthma is well recognised. Baker's asthma and laboratory, animal worker's asthma are seen in atopic individuals, but isocyanate (a paint additive) and other industrial exposures lead to asthma through non-atopic mechanisms.

K. Berg, V. Boulyjenkov, Y. Christen (Eds.)
Genetic Approaches to Noncommunicable Diseases

Asthma may be recognized by questionnaire, physical examination, and the demonstration of variable reduction of airflow. In the absence of an attack of asthma, airflow limitation can be demonstrated by challenge tests. Challenges in common use include exercise, cold air, and inhaled bronchial spasmogens, such as histamine or methacholine. Of these challenges, spasmogen inhalation gives the most reliable measure of underlying airway lability. Challenge tests have been widely used in the investigation of asthma, both in the clinical setting and in large epidemiological surveys.

Of the various types of asthma, atopic asthma is clinically most easily recognised and defined and has the most obvious familial clustering. For this reason most efforts towards elucidating the genetic causes of asthma have been directed at asthma in children and young adults, and at the underlying condition of atopy.

Atopy

Atopy is distinguished by immunoglobulin E (IgE) responses to inhaled proteins, known as allergens. Typical allergen sources include house dust mite (HDM), grass pollens and animal danders (sheddings from skin and fur). The total annual exposure to allergens is small, often in the order of micrograms. IgE binds by its high affinity receptor (FcϵRI), most notably to mast cells in the skin, and in mucosal surfaces of the lung and intestines. Mast cells contain dense granules, which contain histamine and other inflammatory mediators, in addition to pro-inflammatory cytokines. In sensitized individuals, exposure to allergen produces cross-linking of IgE, triggering of high affinity receptors, and release of mast cell granules. The subsequent inflammation occurs in two waves, the first immediate and the second some hours later. Inflammation produces airway narrowing, with wheeze when occurring in the lung, and sneezing and obstruction when in the nose. The regulation of IgE and of some components of early and late inflammation are under the control of antigen-specific T-cells.

The atopic state is detected most easily by skin prick tests. In these, allergen in dilute solution is placed on the skin and a superficial prick is made to introduce minute amounts of allergen below the dermis. Sensitisation and mast cell degranulation are detected by a wheal which is maximal after 10 to 15 minutes. A significant wheal is judged to be between 2 and 4 mm greater than a negative control. Ninety-five percent of individuals who are atopic will react either to HDM or to grass pollen or both.

Atopy may also be detected by elevation of the total serum IgE, or by elevation of serum IgE titres against common allergens. Elevation of antigen-specific IgE is detected by RAST of ELISA techniques. In the affluent populations of the West, there is a close correlation between prick skin tests, specific IgE titres (RASTs), the total serum IgE, and symptoms of wheeze or rhinitis. Despite these close correlations, the relationship between the variables is complex.

Atopy, defined by skin tests, is very common, and has been shown in several large Western population samples to affect between 40% and 50% of young adults (Cline and Burrows 1989; Holford-Strevens et al. 1984; Peat et al. 1987). The prevalence of asthma has risen steadily through this century (Strachan et al. 1994). The prevalence of seasonal rhinitis also appears to have risen, although the evidence for this is less clear. The reasons for this increase, which cannot be due to changes in the gene pool, must be environmental (see below).

Any trait as common as atopy cannot be considered abnormal, and it is obvious that the atopic state gives some advantage to those who carry it. The most likely evolutionary reason for atopy to exist is that IgE is particularly important in handling parasite infestations (Ogilvie and Jones 1969; Capron and Capron 1994). In our society, we are now largely free of parasite infection, the implications of which are also discussed in the section on environemnt below.

Genetic Epidemiology

Factors Confounding Genetic Studies of Atopy

The high population prevalence of atopy may seriously confound genetic studies (Cookson 1994). If atopy is present in 40–50% of the population, then a fifth of marriages may be between two atopics. Any large pedigree is therefore likely to contain several atopy genes introduced through different individuals, instead of a single abnormal gene introduced through one progenitor. If atopy was due to a single gene disorder, then many of the population would be homozygous. If, as is likely, more than one gene predisposes to the syndrome, then many individuals will carry two or more ot these genes.

The substantial prevalence of atopy means that great care has to be taken in the recruitment of families for genetic studies. Ascertainment by public appeals for families with asthma (Moffatt et al. 1992) or with eczema (Coleman et al. 1993) produced samples in which 70% and 80%, respectively, were atopic, with considerable loss of power to detect linkage (Moffatt et al. 1992). For this reason, in Oxford we now recruit families either from population samples (i.e., complete ascertainment) or through a defined proband with atopic disease.

For the purposes of genetic investigations, it is necessary to decide which measures of the atopy or asthma phenotype should be studied. The total serum IgE is an attractive parameter for genetic study, as it has well-established normal values and in large population surveys correlates well with the presence of asthma. However, about 45% of the variation in the total serum IgE is attributable to the specific IgE (RAST) to HDM or grass pollen (Cookson et al. 1991). When multiple regressions are carried out on population data, asthma and bronchial hyper-responsiveness are found to relate to variation in the specific IgE, most notably to HDM (Cookson et al. 1991; Sears et al. 1989). Once specific IgE is taken into account, the residual total IgE does not correlate with the presence of asthma. The specific IgE, either detected indirectly by skin tests or directly in the serum by RAST or ELISA techniques, may therefore be suitable for genetic analysis. Bronchial hyper-responsiveness is a further intermediate phenotype that is currently being investigated.

It is also possible to study asthma as the principal phenotype. If this is done, care needs to be taken that the asthma being investigated is as clinically homogeneous as possible, which will usually mean the asthma of children and young adults. As there is no clear-cut division between normal and abnormal individuals, studies of asthma should exclude marginal phenotypes and concentrate on "barn door" affected and unaffected subjects.

Selection will affect the type of genetic effects found in particular samples of subjects or families. Even if total or specific IgE is used as a phenotype, the factors influencing the IgE in asthmatics may be different form those affecting the IgE in children with eczema, or in subjects selected for the presence of positive skin tests rather than for symptoms.

The behaviour of atopy with age presents a particular problem for geneticists. Whilst many diseases have increasing penetrance throughout life, atopy has a low penetrance in infancy, which rises to a maximum from 15 to 25 years of age. Thereafter the serum IgE and skin prick test responses decline steadily, until at the age of 45 the serum IgE may be half of its value at the age of 15 (Cline and Burrows 1989).

The Inheritance of Atopy

The pattern of inheritance of atopy has been the subject of much debate. In a study of 1000 patients, Cooke and Van der Veer (1916) found that if one parent was allergic, then 50% of the childern were similarly affected; if both parents were allergic, then so too were 75% of their children. This neat Mendelian finding was disputed by Schwartz (1952), who found families in which atopic children had only normal parents; he proposed a dominant model of inheritance tempered with incomplete penetrance. Weiner et al. (1936) observed a similar pattern in 1936.

These early authors studied the inheritance of the whole syndrome, whithout reducing it to its component parts. Later studies concentrated on the inheritance of specific illnesses, such as asthma or hay fever. In these circumstances a pattern of inheritance was much harder to define. Sibbald and Turner-Warwick (1979) studied the first degree relatives of atopic and non-atopic asthmatics, mostly by questionnaire, and found some evidence of familial aggregation of atopy without any clear-cut pattern of inheritance. Edfors-Lubs studied 7000 twin pairs for asthma and atopy, relying primarily on the responses to a questionnaire. She concluded that asthma was polygenic (Edfors-Lubs 1971).

After atopy was shown to be mediated by IgE, researchers concentrated on the genetics of this parameter, as it was quantifiable in a way that was not possible with symptoms alone. Bazaral et al. (1971) studied IgE levels in infants and mothers, concluding that there was simple Mendelian inheritance of basal IgE levels. A subsequent study by the same investigators (Bazaral et al. 1974) showed identical twins to be highly concordant for total serum IgE, and that this effect was not linked to HLA haplotypes. The results indicated an important genetic component to the control of the total serum IgE. Hanson et al. (1985) studied publmonary function, total serum IgE and specific IgE responses (RAST) in mono- and dizygotic twins reared together and apart. He too found that monozygotic twins reared apart or together were concordant for the total IgE, but that they differed in their specific IgE responses.

Marsh and his colleagues (1981) studied many families for the inheritance of total IgE. They found that there was no simple pattern of Mendelian inheritance of the high IgE trait, but that a model in which high IgE was recessive best fitted the data. Gerrard et al. (1974) also studied many families with complex segregation analysis, concluding that a major locus controlled IgE levels, with a recessive allele determining high IgE levels, but that other genes influenced the trait. Blumenthal et al. (1981), however, concluded that a dominant allele coded for high IgE levels in some families and a recessive in others.

Borecki et al. (1985) studied a Canadian population for the inheritance of atopy. They found that if the total serum IgE was the only measure of atopy, then an autosomal recessive pattern of inheritance best fitted the data. When they included symptoms in their definition, they found that a dominant pattern of inheritance best explained their findings.

Using a definition of atopy that included the responses to skin prick tests and serum-specific IgE estimation in addition to the total IgE (which they termed IgE responsiveness), Cookson and Hopkin (1988) examined the genetics of atopy in a limited number of nuclear and extended families. Their findings suggested that there was a major genetic component to atopy. As with the kindreds studied by Marsh, at least 10% of atopic subjects did not have atopic parents, which the authors attributed to dominant inheritance with incomplete penetrance (Cookson and Hopkin 1988).

Thus, diverse models have been proposed for the inheritance of atopy at different times, with varying definitions of atopy and different methods of analysis. It is perhaps as a result of the diversity of approaches that dominant, dominant with incomplete penetrance, recessive and polygenic modes of inheritance have all been suggested. None of these hypotheses explains the results of many studies showing that the risk of atopy is much higher in the children of atopic mothers than in the children of atopic fathers. That asthmatic mothers had more asthmatic children than asthmatic fathers was reported 60 years ago (Bray 1931) and, large studies have shown a similar maternal pattern to the inheritance of elevations of the cord blood IgE (Magnusson 1988; Halonen et al. 1992), atopic symptoms (Arshad et al. 1992; Åberg 1994), and skin prick test responses to common allergens (Kuehr et al. 1993). This finding may be due to interactions between the mother and her child in utero through the placenta, or postpartum through the breast milk. Genomic imprinting, in which a paternal "atopy gene" may be suppressed during spermatogenesis, is also possible (Hall 1990).

One new approach to the problem of a complex phenotype has been the application of regressive models to segregation analysis. Despite the close correlations between symptoms, the total serum IgE, and the specific IgE, genetic effects independently modifying these different variables can be dissected out with these models. This type of segregation analysis has been applied to the total serum IgE. The results demonstrate that the total IgE is influenced by at least one gene that is independent to genes affecting skin tests or positive RAST tests (Dizier et al. 1993).

The failure to show a simple, consistent model of inheritance can most simply be explained by the likelihood that several genes are interacting with a strong environmental component. For the purpose of identifying genes causing atopy, an eclectic approach to phenotype definition is necessary, allowing for potential differences in the genes influencing skin tests and RASTs, the total serum IgE, or disease states such as asthma.

Finding Genes

As with other complex diseases, genes contributing to atopy may be found either by examining candidate genes or by genetic linkage. The most obvious candidate genes for atopy include IL4, γ-interferon, IL10, G-CSF, and the genes making up the high and low affinity receptors for IgE. Also included in this list should be the corresponding ligands or receptors. The enormous increase in understanding of the complex cytokine networks that influence atopy has meant that a plausible case could be put for as many as 20 different candidates. The role of candidate genes may be assessed by defining polymorphisms within the respective genes, and testing for associations with disease. At the moment, a systematic search through the various candidates has not been carried out. Two candidates, IL-4 and the beta chain of the high affinity receptor for IgE ($Fc\epsilon RI\beta$), have been implicated by genetic linkage studies.

Genetic linkage relies on the demonstration of co-inheritance of disease and genetic markers of known chromosomal localisation. This approach has the advantage of not requiring any pre-existing knowledge of the pathophysiology of the disease. However, the power to detect linkage in multigenic diseases is very limited (Table 1); several hundred families may be necessary to detect linkage to a gene affecting a third of subjects with disease. A further problem with complex diseases is that of replication of linkage (Suarez et al. 1994). Linkage to a heterogeneous trait will normally only be found fortuitously, in samples that contain an exceptional proportion of individuals or families influenced by that particular gene. Simulation experiments have shown that, in these circumstances, many studies may be necessary before replication occurs.

Table 1. The power to detect linkage[a]

Fraction linked	Recessive inheritance	Dominant inheritance	Imprinted inheritance
0.80	22	69	29
0.50	62	181	87
0.30	181	508	253
0.20	412	1152	571
0.10	1662	4637	2310

[a]The table shows the number of affected sibling pairs required to detect loci at $p = 0.05$ with 90% power at $\theta = 0.005$, with different proportions of families linked to the putative locus. The required numbers of sib-pairs are estimated for three modes of inheritance. The table assumes 70% marker informativeness. (The table is courtesy of Dr. Alan Young, Statistical Genetics Group, Wellcome Centre for Human Genetic Disease, Oxford OX3 7BN).

Genes InØuencing Asthma and Atopy

A number of genes that influence atopy have now been identified. These may be divided into genes that predispose in general to atopy and those that influence the particular allergens to which atopic individuals react. Genes predisposing to generalised atopy have been identified on chromosome 11 and chromosome 5 by a combination of genetic linkage and candidate gene approaches.

Generalised Atopy

Chromosome 11q 12–13

The first suggested linkage of atopy was to the marker d11s97 on chromosome 11q13 (Cookson et al. 1989; Young et al. 1992). Following some controversy (Marsh and Marsh 1992), this linkage has been replicated by two further groups (Shirakawa et al. 1995a; Colleé et al. 1993). This linkage was confounded by the high prevalence of atopy and because the linkage was predominately seen in maternal meioses (Shirakawa et al. 1995a; Cookson et al. 1992). In the largest study described, linkage was exclusively maternal (Cookson et al. 1992). The reasons for the maternal linkage are not known, and it is not clear that this maternal phenomenon corresponds to the phenotypic maternal inheritance of atopy that has been previously noted.

Recognition of the maternal linkage allowed better localisation of the atopy locus, to within a 7 centiMorgan one lod unit support interval (Sandford et al. 1993, 1995). This interval was centrometric to and excluded the original d11s97 marker to which linkage was first observed. A lymphocyte surface marker, CD20, was noted to be within the interval. CD20 shows sequence homology to the beta chain of the high affinity receptor for IgE (FcϵRIβ) and has been localised close to that gene on mouse chromosome 19 (Hupp et al. 1989). The human Fc∊RIβ was subsequently found to be on chromosome 11q13, in close genetic linkage to atopy (Sandford et al. 1993). Two coding polymorphisms have now been identified within the gene, Fc∊RIβ Leu 181 and Fc∊RIβ Leu181/Leu183 (Shirakawa et al. 1995b). These both show strong associations with atopy when maternally inherited. The population prevalence of Fc∊RIβ Leu181/Leu183 is about 4% (Hiller et al. 1994), and Fc∊RIβ Leu181 has been reported in 15% of asthmatics (Shirikawa et al. 1995b).

These results with Fc∊RIβ variants have not been replicated outside of the Oxford group, and a reliable assay system for the variants has not yet been established. Fc∊RIβ Leu181 does not show functional differences from the wild type receptor (JP Kinet, personal communication). A further complication is the detection of a third homologous gene, Htm4, in close proximity to Fc∊RIβ and CD20 (Adra et al. 1994), so that it is not clear how many members of the gene family are present. Therefore it is not yet established if the chromosome 11q atopy gene is Fc∊RIβ or some other gene in linkage disequilibrium with the Fc∊RIβ variants.

Chromosome 5

Linkage of the total serum IgE to markers near the cytokine cluster on chromosome 5q31–33 was demonstrated by Marsh et al. (1994). Marsh and his colleagues studied Amish pedigrees, selected to contain members with positive skin prick tests. Linkage was strongest in families with the lowest serum IgE. The result was replicated by Myers et al. (1994) in Dutch asthmatic families. Linkage has not been found in other studies of extended families (S. Rich, personal communication). My group have tested 1,500 individuals from 300 nuclear families, and found no evidence for linkage either by sib-pair or by lod score methods. However, N. Vest the claim that linkage is predominantly seen with the low IgE phenotype, we have used class D regressive models to account for the specific IgE response. The residual IgE shows evidence of linkage to a microsatellite repeat found in IL4, but not to the other polymorphic markers studied by Marsh or Myers (Dizier et al., in preparation).

The region contains a number of cytokines, the most important of which, from the point of view of atopy, are IL-4, IL-13, the p40 subunit of IL-12, and IL-5. Other cytokines include IL-9 and granulocyte-colony stimulating factor (G-CSF). A substantial amount of work is now required to establish which of these various candidates accounts for the linkage.

SpeciÆc Atopy

Atopic individuals differ in the particular allergens to which they react. This difference is clinically significant as asthma and bronchial hyperresponsiveness are associated with allergy to house dust mite (HDM) but not grass pollens (Cookson et al. 1991; Sears et al. 1989). It is therefore of interest to examine whether particular genes influence the IgE response to specific allergens. In addition, study of these genes may give an insight into the inheritance of normal variation within the immune system and the functional consequences of such variation.

There are two classes of genes that are likely candidates for constraining specific IgE reactions. These are the genes encoding the human leucocyte antigen (HLA) proteins, and the genes for the T-cell receptor (TCR). These molecules are central to the handling and recognition of foreign antigen.

Inhaled allergen sources such as HDM are complex mixtures of many proteins. A number of "major allergens" to which IgE responses are consistently found in most individuals have been identified from each allergen source. It is likely that genetic associations will be better detected with reactions to purified major allergens, rather than with complex allergens sources. Major allergens include *Der p* I(25.4 kD) and *Der p* II(14.1 kD) from the house dust mite *Dermatophagoides pteronyssinus, Alt a* I(28 kD) from the mould *Alternaria alternata, Can f* I(25 kD) from the dog *Canis familiaris, Fel d* I(18 kD) from the cat *Felis domesticus,* and *Phl p* V(30 kD) from Timothy grass, *Phleum pratense.*

HLA

The human major histocompatibility complex (MHC) includes genes coding for HLA class II molecules (HLA-DR, DQ and DP), which are involved in the recognition and presentation of exogenous peptides.

An HLA influence on the IgE response was first noted by Levine et al. (1972) who found an association between HLA class I haplotypes and IgE responses to antigen E derived from ragweed allergen (*Ambrosia artemisifolia*). This association was subsequently found to be due to restriction of the response to a minor component of ragweed antigen (*Amb a* V) by HLA-DR2 (Marsh et al. 1981). To date the association of *Amb a* V (molecular weight 5,000) and HLA-DR2 is the only HLA association to have been consistently confirmed (Levine et al. 1972; Marsh et al. 1981; Blumenthal et al. 1988). Other suggested associations are of the rye grass antigens Lol p I, Lol p II and Lol p III with HLA-DR3 (in the same 53 allergic subjects; Friednoff et al. 1988; Ansari et al. 1989). American feverfew (*Parthenium hysterophorus*) and HLA-DR3 in 22 subjects from the Indian subcontinent (Sriramarao et al. 1990), the IgE response to Bet v I, the major allergens of birch pollen, and HLA-DR3 in 37 European subjects (Fischer et al. 1992), and an HLA-DR5 association with another ragweed antigen *Amb a* VI in 38 subjects (Marsh et al. 1987).

Other authors have reported negative associations with particular allergens. These include HLA-DR4 and IgE responses to mountain cedar pollen (37 subjects; Reid et al. 1991) and HLA-DR4 and melittin (from bee venom; 22 subjects, Lympany et al. 1990). Non-responsiveness to Japanese cedar pollen may be associated with HLA-DQw8 (Sasazuki et al. 1983).

There is to date no confirmation of many of these results, and the number of subjects has generally not approached that required to establish an unequivocal HLA association. In addition there has not been recognition of the problems of reactivity to multiple allergens; significant relationships between HLA-DR alleles and five antigens (*Amb a* V, Lol p I, Lol p II, Lol p III and *Amb a* 6) have been claimed from the same pool of approximately 200 subjects (Marsh et al. 1981, 1987; Freidhoff et al. 1988; Ansari et al. 1989).

To test more definitively if HLA class II gene products have a general influence on the ability to react to common allergens, we genotyped for HLA-DR and HLA-DP in a large sample of atopic subjects from the British population (Young et al. 1994). The subjects were tested for IgE responses to the most common British major allergens.

A total of 431 subjects from 83 families were genotyped at the HLA-DR and HLA-DP loci and serotyped for IgE responses to six major allergens from common aero-allergen sources. Three hundred subjects were used as controls. The subjects and the controls came from the same relatively homogeneous population. In the United Kingdom and Europe, allergens other than Bet v I and those tested for in our study are uncommon of sensitisation and IgE-mediated allergy.

The results showed only weak associations between HLA-DR allele frequencies and IgE responses to common allergens. A possible excess of HLA-DR1 was found in subjects who were responsive to *Fel-d* I compared to those who were not (Odds Ratio (OR)$=2, p=0.002$), and a possible excess of HLA-DR4 was found in subjects responsive to *Alt a* I(OR$=1.9, p=0.006$). Increased sharing of HLA-DR/DP haplotypes was seen in sibling pairs responding to both allergens. *Der p* I, *Der p* II, *Phl p* V and *Can f* I were not associated with any definite excess of HLA-DR alleles. No significant correlations were seen with HLA-DP genotype and reactivity to any of the allergens.

Of the possible associations, those of *Alt a* I with HLA-DR4 and of *Fel d* I with HLA-DR1 were supported by a finding of excess sharing of a HLA haplotype in affected sibling pairs. Regression analysis showed that the apparent association of *Phl p* V with HLA-DR4 was due to the presence of many individuals who reacted with an IgE response both to *Alt a* I and *Phl p* V. The association of HLA-DR1 and *Fel d* I is the strongest statistically and is significant even taking the multiple comparisons into account.

The study was the first to investigate HLA-DP alleles and reactivity to common allergens. As no definite correlation was found between any antigen response and HLA-DP genotypes with substantial numbers of subjects, HLA-DP genes are unlikely to have a major role in restricting IgE responses to these allergens.

The results suggest that HLA-DR alleles do modify the ability to mount an IgE response to particular antigens. However, the Odds Ratio for the association of *Alt a* I with HLA-DR4 was only 1.9, and that of *Fel d* I with HLA-DR1 was 2.0. Thus class II HLA restriction seems insufficient to account for individual differences in reactivity to common allergens. It is therefore likely that environmental factors or other loci, such as T-cell receptor genes, may be of greater relevance in determining an individual's susceptibility to specific allergens.

The T-cell Receptor (TCR)

The T-cell receptor is usually made up of α and β chains, although 5% of receptors consist of γ and δ chain locus is on chromosome 7, and the α chain locus is on chromosome 14. The δ chain genes are found within the α chain locus.

An enormous potential for TCR variety follows from the presence of many variable (V) and junctional (J) segments within the TCR loci. However, the usage of the TCR Vα and Vβ segments by lymphocytes is not random and may be under genetic control (Loveridge et al. 1991; Moss et al. 1993; Gulwani-Akolar et al. 1991; Robinson 1992).

To examine if the TCR genes influence susceptibility to particular allergenes, we tested for genetic linkage between IgE responses and microsatellites from the TCR-α/δ and TCR-β regions (Moffatt et al. 1994). Two independent sets of families, one British and one Australian, were investigated. Because the mode of inheritance was unknown, and because of interactions from the environment and other loci, affected sibling pair methods were used to test for linkage.

No linkage of IgE serotypes to TCR-β was detected, but significant linkage of IgE responses to the house dust mite allergens *Der p* I and *Der p* II , the cat allergen *Fel d* I, and the total serum IgE to TCR-α was seen in both family groups. The results show that a locus in the TCR α/δ region is modulating IgE responses. The close correlation between total and specific IgE makes it difficult to determine if the locus controls specific IgE reactions to particular allergens or confers generalised IgE responsiveness. Nevertheless, linkage was strongest with highly purified allergens, suggesting that the locus primarily influences specific responses. The pattern of allele sharing seen with some serotypes suggests a recessive genetic effect, making it possible that this linkage corresponds to the recessive atopy locus implied by previous segregation analyses (Dizier et al. 1993; Gerrard et al. 1978).

Replication of positive results of linkage in a second set of subjects is important in interpreting this study. Differences between the populations for the serotypes showing TCR-α allele sharing may be due to different allergen exposures, as grass pollen responses were much more common in Australian subjects. In addition, British subjects were recruited through clinics, whereas Australian subjects were not selected by symptoms.

No association was seen between particular IgE responses and specific TCR-α microsatellite alleles, implying that the microsatellite is not in immediate proximity to the IgE modulating elements. The degree of linkage disequilibrium across the TCR-α/δ locus seems low (Robinson and Kindt 1987), and the microsatellite has only been localised within a 900 kb yeast artificial chromosome (Cornélis et al. 1992). The observed linkage may therefore be with any elements of TCR-α or TCR-δ, or with other genes in the locality.

Several Vα genes have been recognised to be polymorphic (Cornélis et al. 1993), and limitation of the response to an allergen may correspond to these polymorphisms. Particular TCR-Vα usage may induce IL-4 dominant (Th2) helper T-cells, which enhance IgE production (Heinzel et al. 1991). A reported non-random usage of Vα13 usage in *Lol p* I specific T-cell clones supports independently the possibility of Vα genes controlling IgE responses (Mohapatra et al. 1994).

The TCR-δ locus is also a candidate for this linkage. The function of TCR-γ/δ cells is not known, but their location on mucosal surfaces, where allergens initiate IgE responses, could suggest a role in IgE regulation (Holt and Mc Menamin 1991).

This study has therefore identified a further genetic locus affecting atopy. The genetic restriction of specific IgE responses may be clinically significant and may be of general interest in understanding the control of humeral immunity. Further localisation of this genetic effect requires the identification of TCR α/δ elements showing allelic associations with specific IgE responses. Studies are also needed to investigate the interactions between this chromosome 14 linkage and the HLA class II genes.

A Genome Screen for Atopy and Asthma

The four loci described above do not account for all atopy. The chromosome 11 and 5 genes do not seem to have major effects on the population as a whole, and HLA and TCR-α loci modify the specific response rather than endowing any general predisposition to atopy. Segregation analysis is unable to predict with any accuracy the number and nature of genes contributing to atopy and asthma. To discover if atopy is a genuine polygenic disorder, my group have carried out a complete genome screen in 80 nuclear families, with 260 markers spaced at approximately 10% recombination. Using sib-pair analysis we have discovered three probable new linkages ($p < 0.001$) and three further possible linkages ($p < 0.01$). Linkage at most of these loci is also seen to the asthma phenotype. Before releasing these localisations, we are attempting to confirm each linkage in 300 further asthmatic families. We intend to complete a screen in a final total of 250 families, in three groups ascertained for atopy, asthma and eczema. Similar large-scale genome scans are to be carried out in the United States and Canada, so that it is likely that general agreement will soon be reached on the number and nature of the most important loci causing atopy.

Genes and Environment

No description of the genetics of asthma would be complete without some consideration of the effects of environment. In the absence of environmental precipitants, allergic asthma and hay fever would not exist. Such conditions are found on mountains, where there is little pollen, and where low humidity prevents house dust mite growth. Schoolchildren raised at high altitude develop less allergy than those raised at sea level (Charpin et al. 1991). Similarly, children living in the dry interior of Australia develop less allergy than those living in more humid conditions near the coast (Peat et al. 1987).

Data from a number of sources indicate that events in early infancy are critical in determining the subsequent course of allergic disease. In the Scandinavian countries a short intense spring flowering of birch trees is accompanied by symptoms in many individuals. Children born in the three months around the pollen season carry an increased risk of allergy to birch pollen for the rest of their life (Holt et al. 1990). In English children the level of house dust mite in infants' bedding during the first year of life correlates with the subsequent risk of childhood asthma (Sporik et al. 1990).

The enormous increase in the prevalence of asthma in the past two decades cannot be attributed to changes in gene frequencies in the affected populations, and must be due to an environmental factor or factors. Air pollution has been suggested as a cause of this increase, although atmospheric pollution has declined steadily since the 1950's in England and Western Europe, and ozone levels remained stable despite an increase in the number of cars. Comparative studies of the prevalence of asthma have been carried out between East and West Germany, two regions with genetically similar populations, but with far higher levels of atmospheric pollution in the East (von Mutius et al. 1992). Surprisingly, the prevalence of asthma is lower in the East than in the West. This decrease is matched by a lower prevalence of atopy, as detected by skin tests to common allergens (von Mutius et al. 1994a). Similar results are seen when the prevalence of asthma in the Baltic States is compared to that of Sweden (Bråbäck et al. 1994). This difference may be attributable to childhood respiratory infections, as pollution and overcrowding, both of which are more common in the East, predispose to infantile infection. Support for this hypothesis is given by the finding that the youngest children in large sibships have significantly less asthma than their older siblings (von Mutius et al. 1994b). At the cellular level it is suggested that early infections program the immature immune system towards a Th2 rather than a Th1 helper cell profile, thereafter favouring a cellular rather than humeral immune response.

Thus, even in genetically similar people, the dose and timing of allergen exposure will have important effects on subsequent manifestations of the atopy phenotype. This places an additional requirement for careful study design and interpretation in attempts to unravel the genetics of atopic disease.

Another environmental factor to consider is that of parasitism. Slum-dwelling Venezuelan children have higher levels of serum IgE and lower levels of asthma than their more affluent (Lynch et al. 1993). In endemically parasitised Australian Aborigines the presence of a positive RAST to HDM correlates poorly both with skin test responses to the same allergen and the presence of asthma. Multiple regression shows that the discrepancy between RAST and skin tests is accountable by the elevation of total serum IgE. The results fit the hypothesis that parasitism, by causing an increase in polyclonal IgE, is protective against atopy.

Screening

The increase in prevalence of asthma in recent decades has an important corollary: asthma may be preventable. Recognition of children or infants genetically predisposed to asthma is likely to be the first step in strategies for prevention by environmental manipulation or vaccination in the first year of life. At the moment, the high affinity IgE receptor variant FcϵRIβ Leu 181/Leu 183 appears to be a significant risk factor for atopy and for bronchial hyper-responsiveness, but can only account for a small proportion of atopy and asthma in the general population. It is likely that other genes will be identified in the near future, and that a comprehensive estimation of genetic risk to a particular infant will be feasible by the end of the decade.

References

Aberg N (1994) Familial occurence of atopic disease: genetic versus environmental factors. Clin Exp Allergy 23: 829–834
Adra CN, Lelias J-M, Kobayashi H, Kaghad M, Morrison P, Rowley JD, Lim B (1994) Cloning of the cDNA for a haemopoietic cell-specific protein related to CD20 and the beta subunit of the high-affinity IgE receptor: Evidence for a family of proteins with four membrane spanning regions. Proc Natl Acad Sci 91: 10178–10182
Ansari AA, Freidhoff LR, Meyers DA, Bias WB, Marsh DG (1989) Human immune responsiveness to Lolium perenne pollen allergen Lol p III (rye III) is associated with HLA-DR3 and DR5 [published erratum appears in Human Immunol 26: 149]. Human Immunol 25: 59–71
Arshad SH, Matthews S, Grant C, Hide DW (1992) Effect of allergen avoidance on development of allergic disorders in infancy. Lancet 339: 1493–1497
Bazaral M, Orgel HA, Hamburger RN (1971) IgE levels in normal infants and mothers and an inheritance hypothesis. J Immunol 107: 794–801
Bazaral M, Orgel HA Hamburger RN (1974) Genetics of IgE and allergy: serum IgE levels in twins. J Allergy Clin Immunol 54: 288–304
Blumenthal MN, Namboodiri K, Mendell N, Gleich G, Elston RC, Yunis E (1981) Genetic transmission of serum IgE levels. Am J Med Genet 10: 219–228
Blumenthal MN, Johnson B, Marcus D, Alper C, Mendell N, Thode H, Yunis E (1988) Immune response genes of ragweed sensitive individuals. J Allergy Clin Immunol 81: 307
Boreki I, Rao DC, Lalovel JM (1985) Demonstration of a common major gene with pleiotrophic effects on Immunoglobulin E and allergy. Genet Epidemiol 2: 327–328
Bråbäck L, Breborowicz A, Dreborg S, Knutsson A, Pieklik H, Björkstén B (1994) Atopic sensitization and respiratory symptoms among Polish and Swedish school children. Clin Exp Allergy 24: 826–835
Bray GW (1931) The hereditary factor in hypersensitiveness anaphlaxis and allergy. J Allergy II: 205–224

Capron M, Capron A (1994) Immunoglobulin E and effector cells in schistosomiasis. Science 264: 1876-1877

Charpin D, Birnbaum J, Haddi E, Genard G, Lanteaume A, Toumi M, Faraj F, Van der Brempt X, Vervolet D (1991) Altitude and allergy to house dust mites. A paradigm of the influence of environmental exposure on allergic sensitisation. Am Rev Resp Dis 143: 983-986

Cline MG, Burrows BB (1989) Distribution of allergy in a population sample residing in Tuscon, Arizona. Thorax 44: 425-432

Coleman R, Trembah RC, Harper JI (1993) Chromosome 11q13 and atopy underlying atopic eczema. Lancet 341: 1121-1122

Collée JM, ten Kate LP, de Vries HG, Kliphuis JW, Bouman K, Scheffer H, Gerritsen J (1993) Allele sharing on chromosome 11q13 in sibs with asthma and atopy. Lancet 342: 936

Cooke RA, van der Veer A (1916) Human sensitisation. J Immunol 1: 201

Cookson WOCM (1994) Atopy: A complex genetic disease. Ann Med 26: 351-353

Cookson WOCM, Hopkin JM (1988) Dominant inheritance of atopic immunoglobulin-E responsiveness. Lancet i: 86-88

Cookson WOCM, Sharp PA, Faux JA, Hopkin JM (1989) Linkage between immunoglobulin E responses underlying asthma and rhinitis and chromosome 11q. Lancet i: 1292-1295

Cookson WOCM, De Klerk NH, Ryan GR, James AL, Musk AW (1991) Relative risks of bronchial hyper-responsiveness associated with skin-prick test responses to common antigens in young adults. Clin Exp Allergy 21: 473-479

Cookson WOCM, Young RP, Sandford AJ, Moffatt MF, Shirakawa I, Sharp PA, Faux JA, Julier C, Le Souef PN, Nalaumura Y, Lathrop GM, Hopkin JM (1992) Maternal inheritance of atopic IgE responsiveness on chromosome 11q. Lancet 340: 381-384

Cornélis F, Hashimoto L, Loveridge J, MacCarthy A, Buckle V, Julier C, Bell J (1992) Identification of a CA repeat at the TCRA locus using yeast artificial chromosomes: a general method for generating highly polymorphic markers at chosen loci. Genomics 13: 820-825

Cornélis F, Pile K, Loveridge J, Moss P, Harding C, Julier C, Bell JI (1993) Systematic study of human $\alpha\beta$ T-cell receptor V segments shows allelic variations resulting in a large number of distinct TCR haplotypes. Eur J Immunol 23: 1277-1283

Dizier MH, Hill M, James A, Faux J, Ryan G, le Souef P, Musk AW, Lathrop M, Demenais F, Cookson W (1993) Genetic control of IgE level after accounting for specific atopy. Genet Epidemiol 10: 333-334

Edfors-Lubs ML (1971) Allergy in 7000 twin pairs. Acta Allergol 26: 249-285

Fischer DF, Pickl WF, Fae I, Ebner C, Ferreira F, Breiteneder H, Vikoukal E, Scheiner O, Kraft D (1992) Association between IgE response against Bet v I, the major allergen of birch pollen, and HLA-DRB alleles. Human Immunol 33: 259-265

Freidhoff LR, Ehrlich-Kautzky E, Meyers DA, Ansari AA, Bias WB, Marsh DG (1988) Association of HLA-DR3 with human immune response to Lol p I and Lol p II allergens in allergic subjects. Tissue Antigens 31: 211-219

Gerrard JW, Horne S, Vickers P, Mac Kenzie JW, Goluboff N, Garson JZ, Maningasc S (1974) Serum IgE levels in parents and children. J Pediatr 85: 660-663

Gerrard JW, Rao DC, Morton NE (1978) A genetic study of immunoglobulin E. Am J Human Genet 30: 46-58

Gulwani-Akolar B, Posnett DN, Janson CH, Grunewald J, Wigzell H, Akolkar P, Gregersen PK, Silver J (1991) T cell receptor V-segment frequencies in peripheral blood T cells correlate with human leukocyte antigen type. J Exp Med 174: 1139-1146

Hall JG (1990) Genomic imprinting. Arch Dis Childhood 65: 1013-1016

Halonen M, Stern D, Taussig LM, Wright A, Ray CG, Martinez FD (1992) The predictive relationship between serum IgE levels at birth and subsequent incidences of lower respiratory illnesses and eczema in infants. Am Rev Resp Dis 146: 866-870

Hanson B, Kronenberg R, Johnson B, Blumenthal M (1985) Pulmonary function, serum IgE levels and specific IgE responses in monozygotic twins reared apart. J Allergy Clin Immunol 75: 155

Heinzel FP, Sadic MD, Mutha SS, Locksley RM (1991) Production of interferon gamma, interleukin 2, interleukin 4, and interleukin 10 by CD4+lymphocytes in vivo during healing and progressive murine leishmaniasis. Proc Natl Acad Sci USA 88: 7011-7015

Hill MR, Daniels SE, James AL, Faux JA, Ryan G, le Souef P, Musk AW, Cookson WOCM (1994) A coding polymorphism for FcεRIβ in a random population sample: associations with atopy. Genet Epidemiol 11: 297

Holford-Strevens V, Warren P, Wong C, Manfreda J (1984) Serum total immunoglobulin E levels in Canadian adults. J Allergy Clin Immunol 73: 516-522

Holt PG, McMenamin C (1991) IgE and mucosal immunity: studies on the role of intraepithelial Ia+ dendritic cells and δ/γ T-lymphocytes in regulation of T-cell activation in the lung. Clin Exp Allergy 21(Suppl):148-152

Holt PG, McMenamin C, Nelson D (1990) Primary sensitisation to inhalant allergens during infancy. Pediatr Allergy Immunol 1: 3-15

Hupp K, Siwarski D, Mock BA, Kinet JP (1989) Gene mapping of the three subunits of the high affinity FcR for IgE to mouse chromosomes 1 and 19. J Immunol 143: 3787-3791

Kuehr J, Karmaus W, Forster J, Frischer T, Hendel-Kramer A, Moseler M, Stephan V, Urbanik R, Weiss F (1993) Sensitisation to four common inhalant allergens within 302 nuclear families. Clin Exp Allergy 23: 600-605

Levine BB, Stember RH, Rontino M (1972) Ragweed hayfever: genetic control and linkage to HL-A haplotypes. Science 178: 1201-1203

Loveridge JA, Rosenberg WMC, Kirkwood TBL, Bell JI (1991) The genetic contribution to human T-cell receptor repertoire. Immunology 74: 246-250

Lympany P, Kemeny DM, Welsh KI, Lee TH (1990) An HLA-associated nonresponsiveness to mellitin: a component of bee venom. J Allergy Clin Immunol 86: 160-170

Lynch NR, Hagel I, Perez M, Di Prisco MC, Lopez R, Alvarez N (1993) Effect of antihelmintic treatment on the allergic reactivity of childen in a tropical slum. J Allergy Clin Immunol 92: 404-411

Magnusson CG (1988) Cord serum IgE in relation to family history and as predictor of atopic disease in early infancy. Allergy 43: 241-251

Marsh DG, Myers DA (1992) A major gene for allergy-fact or fancy? Nature Genet 2: 252-254

Marsh DG, Meyers DA, Bias WB (1981) The epidemiology and genetics of atopic allergy. New Engl J Med 305: 1551-1559

Marsh DG, Freidhoff LR, Ehrlich-Kautzky E, Bias WB, Roebber M (1987) Immune responsiveness to Ambrosia artemisiifolia (short ragweed) pollen allergen Amb a VI (Ra6) is associated with HLA-DR5 in allergic humans. Immunogenetics 26: 230-236

Marsh DG, Neely JD, Breazeale DR, Ghosh B, Freidhoff LR, Erlich-Kautzky E, Schou C, Krishnaswamy G, Beaty TH (1994) Linkage analysis of IL4 and other chromosome 5q31.1 markers and total serum IgE concentrations. Science 264: 1152-1155

Moffatt MF, Sharp PA, Faux JA, Young RP, Cookson WOCM, Hopkin JM (1992) Factors confounding genetic linkage between atopy and chromosome 11q. Clin Exp Allergy 22:1046-1051

Moffatt MF, Hill MR, Cornelis F, Schou C, Faux JA, Young RP, James Al, Ryan G, le Souef P, Musk AW, Hopkin JM, Cookson WOCM (1994) Genetic linkage of the TCR-α/δ region to specific Immunoglobulin E responses. Lancet 343:1597-1600

Mohapatra SS, Mohapatra S, Yang M, Ansari AA, Parronchi P, Maggi E, Romagnani S (1994) Molecular basis of cross-reactivity among allergen-specific human T Cells. T-cell receptor Vα gene usage and epitope structure. Immunology 81:15-20

Moss PAH, Rosenberg WMC, Zintzaras E, Bell JI (1993) Characterization of the human T cell receptor α-chain repertoire and demonstration of a genetic influence on Vα usage. Eur J Immunol. 23:1153–1159

Myers DA, Postma DS, Panhuysen CIM, Xu J, Amelung PJ, Levitt RC, Bleeker ER (1994) Evidence for a locus regulating total serum IgE levels mapping to chromosome 5. Genomics 23:464–470

Ogilvie BM, Jones VE (1969) Protective immunity in helminth diseases. Proc Royal Soc Med 62: 298–301

Peat JK, Britton WJ, Salome CM, Woolcock AJ (1987) Bronchial hyperresponsiveness in two populations of Australian school children III. Effect of exposure to environmental allergens. Clin Allergy 17: 271–281

Reid MJ, Nish WA, Whisman BA, Goetz DW, Hylander RD, Parker WA Jr, Freeman TM (1992) HLA-DR4-associated nonresponsiveness to mountain cedar allergen. J Allergy Clin Immunol. 89: 593–598

Robinson MA (1992) Usage of human T-cell receptor V beta, J beta, C beta and V alpha gene segments is not proportional to gene number. Human Immunol 35: 60–67

Robinson MA, Kindt TJ (1987) Genetic recombination within the human T-cell receptor alpha-chain complex. Proc Natl Acad Sci USA 84: 9098–9093

Sandford AJ, Shirakawa T, Moffatt MF, Daniels SE, Ra C, Faux JA, Young RP, Nakamura Y, Lathrop GM, Cookson WOCM, Hopkin JM (1993) Localisaiton of atopy and the β subunit of the high affinity IgE receptor (FcεRI) on chromosome 11q. Lancet 341: 332–334

Sandford AJ, Moffatt MF, Daniels SE, Nakamura Y, Lathrop GM, Hopkin JM, Cookson WOCM (1995) A genetic map of chromosome 11q, including the atopy locus. Eur J Human Genet, in press

Sasazuki T, Nishimura Y, Muto M, Ohta N (1983) HLA-linked genes controlling immune response and disease susceptibility. Immunol Rev 70: 51–75

Schwartz M (1952) Heredity in bronchial asthma. Acta Allergol 5 (suppl 2)

Sears MR, Herbison GP, Holdaway MD, Hewitt CJ, Flannery EM, Silva PA (1989) The relative risks of sensitivity to grass pollen, house dust mite and cat dander in the development of childhood asthma. Clin Allergy 18: 419–424

Shirakawa T, Morimoto K, Hashimoto T, Furuyama J, Yamamoto M, Takai S (1995a) Linkage between severe atopy and chromosome 11q in Japanese families. Clinical Genetics, in press

Shirakawa TS, Li A, Dubowitz M, Dekker JW, Shaw AE, Faux JA, Ra C, Cookson WOCM, Hopkin JM (1995b) Association between atopy and variants of the β subunit of the high-affinity immunoglobulin E receptor. Nature Genet, in press

Sibbald B, Turner-Warwick M (1979) Factors influencing the prevalence of asthma in first degree relatives of extrinsic and intrinsic asthmatics. Thorax 34: 332–337

Sporik R, Holgate S, Platts-Mills TAE, Cogswells JJ (1990) Exposure to house dust mite allergen der P1 and the development of asthma in children. New Engl J Med 323: 502–507

Sriramarao P, Selvakumar B, Damodaran C, Rao BS, Prakash O, Rao PV (1990) Immediate hypersensitivity to Parthenium hysterophorus I. Association of HLA antigens and Parthenium rhinitis. Clin Exp Allergy 20: 555–560

Strachan DP, Anderzon HR, Limb ES, O'Neill A, Wells N (1994) A national survey of asthma prevalence, severity, and treatment in Great Britain. Arch Dis Childhood 70: 174–178

Suarez BK, Hampe CL, Van Eerdewegh P (1994) Problems of replicating linkage claims in psychiatry In: Gershon ES, Cloninger CR (eds) Genetic approaches to mental disorders. American Psychiatric Press Inc, Washington 23–46

von Mutius E, Fritzsch C, Weiland SK, Roell G, Magnussen H (1992) Prevalence of asthma and allergic disorders among children in united Germany: a descriptive comparison. Brit Med J 305: 1395–1399

von Mutius E, Martinez FD, Fritzsch C, Nicolai T, Roell G, Thiemann HH (1994a) Prevalence of asthma and atopy in two areas of West and East Germany. Am J Respir Crit Care Med 149: 358–364
von Mutius E, Martinez FD, Fritzsch C, Nicolai T, Reitmer P, Thiemann HH (1994b) Skin test reactivity and number of siblings. Brit Med J 308: 692–695
Weiner A, Zieve I, Fries J (1936) The inheritance of allergic diseases. Ann Eugen 7: 141
Young RP, Lynch J, Sharp PA, Faux JA, Cookson WOCM, Hopkin JM (1992) Confirmation of genetic linkage between atopic IgE responses and chromosome 11q13. J Med Genet 29: 236–238
Young RP, Dekker JW, Wordsworth BP, Schou C, Pile KD, Matthiesen F, Rosenberg WMC, Bell JI, Hopkin JM, Cookson WOCM (1994) HLA-DR and HLA-DP genotypes and immunoglobulin E responses to common major allergens. Clin Exp Allergy 24: 431–439

Prospects of Cancer Control Through Genetics

J.J. MULVIHILL

Summary

At the level of the cell, cancer is a genetic disease; in populations, most types of cancer have some genetic, congenital, or familial determinants. Virtually all cancers have somatic chromosomal abnormalities, and a few constitutional cytogenetic defects predispose to malignancy. At least 338 single gene traits have neoplasia as a feature or complication. Familial syndromes of cancer are being delineated; some have clear genetic origins, others are probably environmentally induced. A 1986 workshop at the US National Institutes of Health (Fogarty International Center) proposed strategies for controlling cancer through genetics in four areas: clinical practice, educational and administrative measures, research needs, and ethical issues. Further deliberation is needed to expand the Workshop proposal to a global view, but feasibility demonstrations are well underway. An international approach to familial cancer being launched by the UICC (the worldwide federation of voluntary national cancer societies) may help catalyze professional and public attention to cancer control through genetics.

In its broadest sense "cancer control" includes primary prevention (blocking the mechanisms of cellular initiation and tumor promotion that result in a clinically recognized cancer), secondary prevention (screening for early clinical diagnosis), and tertiary measures, such as treatment, continuing care, and rehabilitation.

For success in the primary prevention of cancer, it is usually necessary but often not sufficient to understand its etiology. Each cancer probably arises from a complex interaction of environmental exposures with variable susceptibilities of the host (Mulvihill 1993). Even with knowledge of the causal association between some environmental agents and human cancer, efforts for prevention have been thwarted, sometimes by the lack of governmental control and taxation, sometimes by the successes of modern advertisers and the failures of health educators, and perhaps by the grip of human habits.

At the level of the cell, cancer is a genetic disease (Cavenee and White 1995); for, if the cancer cell did not pass on its new rules for escaping the good behavior of normal cells, then no one would die of a clinical cancer. This

K. Berg, V. Boulyjenkov, Y. Christen (Eds.)
Genetic Approaches to Noncommunicable Diseases

commentary is based on the opinion that more could be done, with present knowledge, than is being done to control cancer through genetics. After a brief review of the clinical genetics of cancer, suggested guidelines for cancer control through genetics are offered, and an assessment is provided of the accomplishments since a 1986 workshop on the topic.

Genetic Determinants for Cancer

The genetics of cancer can be considered in three categories: cytogenetics, Mendelian (single gene) traits, and familial or multifactorial inheritance. Long and short reviews are available (Chaganti and German 1985; Lynch and Lynch 1985, 1994; Mitelman 1991; Müller and Weber 1985; Mulvihill et al. 1977; Mulvihill 1989, 1993; Philippe 1989; Weber et al. 1995); the following summation selectively illustrates the broad range of current information and the promise of future progress.

Cytogenetics

With every technical development, Theodor Boveri's 1914 hypothesis gains support: probably every cancer has abnormal chromosomes (Boveri 1914; Mitelman 1991; Rabbits 1994). Most often the cytogenetic defect is somatic only, confined to the malignant cell line or closely related tissue. But, cancer is a complication of a few recognized syndromes of multiple malformations that have underlying constitutional chromosomal defects (Mulvihill 1993). Such cytogenetic syndromes include: Down syndrome (trisomy 21) with various acute leukemias and perhaps testicular cancer; Klinefelter syndrome (47, XXY) with nongonadal germ cell tumors and breast cancer; gonadal dysgenesis, including the full Turner syndrome (with gonadoblastoma if some Y chromosome material is present), mosaic trisomy 8 with preleukemia; the Miller syndrome of aniridia with Wilms tumor and a deletion of 11q13; retinoblastoma with or without birth defects and a deletion of 13q14; and the fragile X syndrome with unusual cancers, including testicular cancer.

Rare familial aggregations of specific cancers have been explained by constitutional translocations: t(3;8) in one family with renal cell carcinoma, an insertion of 11q13 into chromosome 2q32 with familial Wilms tumor, an insertion of 13q14 into 3p12 with familial retinoblastoma, and a t(14;22) with familial meningioma. Recognition of these rare families provides excellent opportunities for clinicians to make presymptomatic or prenatal diagnoses by vigorous identification and screening of persons at high risk. These families also present opportunities for cancer biologists.

The much longer list of *acquired* cytogenetic defects includes many leukemias and lymphomas and some solid tumors with nonrandom abnormalities (Mitelman 1991). Eventually these abnormalities should produce routes to secondary prevention and control. The eleventh edition of *Mendelian Inheritance in Man* (McKusick 1994) enumerates 6678 human traits with definite or suggestive evidence of Mendelian inheritance behavior. At least 338 conditions (in addition to 56 protooncogenes) have neoplasia as the sole feature, a frequent concomitant, or a rare complication (Mulvihill 1993, 1995). The inference can be drawn that 6% of known human genes influence the expression of neoplasia.

Some 22 traits are clear examples of ecogenetics in cancer; that is, each represents an inborn susceptibility to environmental agents that results or predisposes to cancer (Mulvihill 1994). Examples include: xeroderma pigmentosum, cutaneous albinism, and the dysplastic nevus syndrome with unusual sensitivity to ultraviolet radiation resulting in skin cancers; the X-linked hyperproliferative syndrome of Purtillo with Burkitt and other lymphomas arising from unusual susceptibility to Epstein-Barr virus; and hemochromatosis and tyrosinemia with hepatocellular carcinoma following dietary exposure to iron and tyrosine, respectively.

Chromosomal assignment is known of the some 85 (25%) of the 338 preneoplastic traits. Tumor karyotyping can suggest candidate areas of the genome for mapping genes that contribute to neoplasia. As the protein products of cancer genes become known, novel routes to cancer control and prevention are likely to emerge, including specific antibodies, vaccines, antisense and triple helix oligonucleotides, and gene transfer therapy.

Familial Cancer

Since one in four North Americans and Europeans gets cancer in a lifetime, most people will have some relatives with cancer and some, by chance, will have many. In two clinical surveys, 6% of persons with cancer said they had three or more first-degree relatives with cancer (Albano et al. 1981; Müller 1985). In a study that independently confirmed with medical records the accuracy of reported cancers in relatives, the primary site of cancer was correct in 83% of first-degree, 67% of second-degree, and 60% of third-degree relatives (Love et al. 1985).

No firm definition of a cancer family is available so, for now, it must be operational (Mulvihill 1985; Weber et al. 1995), depending on the type and site of cancer, the age at diagnosis, the sex and the numbers of tumors, as well as the absolute numbers of affected relatives. There are many reports of familial recurrences of exactly the same cancer – for example, of the breast, colorectum, ovary or lung – and of closely related tumors, such as squamous carcinomas of the lung and larynx, or adenocarcinomas of the breast and ovary. Some kindreds of apparently unrelated malignant neoplasms are especially provocative and have a known molecular basis.

Two distinctive patterns of aggregation have earned the label "cancer family syndrome," although many others surely remain to be delineated. One is the cancer family syndrome of Lynch (Lynch and Lynch 1985), which is characterized by two or more generations with cancer of the colon and endometrium at an early age of diagnosis and with an excess of persons with multiple primary cancers. An alternate name is Lynch syndrome 2 or hereditary nonpolypotic colonic cancer 2 (HNPCC2); the term Lynch syndrome 1 (HNPCC1) is reserved for familial aggregation of just colon cancers without polyposis or other cancers. The Lynch 2 syndrome has constitutional mutations of genes known by their homologues in yeast to control cell cycling and mitosis. The genes are hMSH2, hMLH1, PMS1, and PMS2. The second pattern of aggregation is the Li-Fraumeni cancer family syndrome or SBLA syndrome as an acronym for the tumor types Sarcomas, Breast, Bone, and Brain tumors, Lung, Laryngeal, and Leukemia and Adrenal cortical neoplasia (Li et al. 1988). Many but not all persons with the Li-Fraumeni syndrome have constitutional mutations of the TP53 gene, whose protein product, p53, is likewise pivotal in the regulation of cell division (Malkin 1994).

NIH Workshop Recommendations

In January 1986, the Fogarty Institutional Center of the US National Institutes of Health (NIH) convened 29 biomedical scientists for a workshop on "Strategies for Controlling Cancer through Genetics" (Parry et al. 1987a,b). In assembling experts in clinical genetics, cancer research, and cancer control, the working premise was that the experience of applying knowledge of medical genetics to the control of congenital and hereditary non-neoplastic diseases should provide models and strategies for cancer control through genetics. Geneticists have learned that the route from knowledge of etiology to control and prevention includes many steps: professional and public education and consensus building, motivation for legislation and enforcement (if prevention is societal), and individual counseling and provision of health care (if prevention is based on personal action). In contrast to some efforts for cancer control, medical geneticists have started with a firm foundation of indisputable scientific facts, undertaking small pilot efforts (in part because there are many more distinct genetic diseases than separate cancers), informing the target population and gaining their assent and endorsement, and conducting several efforts at the same time, e.g., the newborn period for screening for several inborn errors of metabolism.

The workshop's recommendations addressed four areas: improvements of clinical practice, educational and administrative measures, research needs, and ethical issues (Table 1).

Subsequent Developments

What has taken place in the nine years since the NIH Workshop? The brief answer might be, "More than we expected, less than we would like."

In the research arena, the role of mutant genes in the origins and pathogenesis of virtually all cancers has continued to be documented in both targeted and

Table 1. Recommendations of 1986 National Institutes of Health workshop on strategies for controlling cancer through genetics[a]

I. Clinical guidelines: Identifying individuals and families for possible genetic evaluation
   A. *Individuals* with a cancer that is:
      1. bilateral, as separate primaries,
      2. multifocal, within one organ,
      3. an additional primary malignancy,
      4. at an atypical age,
      5. at an atypical site,
      6. in the sex not usually affected,
      7. associated with birth defects,
      8. associated with a Mendelian trait,
      9. associated with a precursor lesion,
      10. associated with a rare disease,
      11. a rare or unusual tumor type.
   B. *Families* with:
      1. one first-degree relative[b] with a cancer with any of the above features,
      2. two first-degree relatives[b] with any cancer.

II. Selected administrative and educational measures to promote the guidelines:
   A. encourage cancer clinicians to recognize high-risk individuals and families,
   B. educate medical geneticists in practice and in training in cancer genetics,
   C. educate the public through communications media and telephone hotlines,
   D. identify centers with specialized knowledge,
   E. add geneticists as consultants to cooperative clinical trials groups.

III. Research needs:
   A. foster interdisciplinary cancer research projects,
   B. establish regional repositories for storing DNA and tumors from selected patients,
   C. collect population-based data on the frequency of genetic and familial cancers,
   D. continue investigations of genetic mechanisms in oncogenesis, e.g., chromosomal rearrangements, oncogene activation, loss of heterozygosity, DNA repair pathways,
   E. continue mapping and cloning cancer-predisposing genes,
   F. quantitate genetic risk factors in cancer, perhaps with twin studies,
   G. explore genetic determinants of environmentally induced cancers and adverse responses to therapy,
   H. establish quality control procedures for genetic laboratory methods, e.g., cytogenetics, DNA methods,
   I. develop user-friendly, multiple-access supercomputers for biomedical research.

IV. Issues in ethics:
   A. protect individuals' and families' privacy,
   B. protect confidentiality of records and tests,
   C. protect insurability,
   D. reach out to family members at high risk,
   E. prevent the misuse of predictive tests,
   F. accept directive counseling contrary to traditional genetic counseling.

[a]Mulvihill 1989; Parry et al. 1987a,b.
[b]Brother, sister, mother, father, or child.

basic research (Cavenee 1995). Among the high points, one can cite: the molecular elucidations of some of the nonrandom cytogenetic findings in all cancers (Rabbits 1994); the mapping and sometimes cloning of genes that predispose to even the commonest cancers of human beings (e.g., BRCA1, BRCA2, MSH2, MLH1, PMS1, PMS2, TP53, NF1, NF2, the nevoid basal cell carcinoma syndrome, and multiple endocrine neoplasia 1 and 2); the increasing relevance of years of basic research, often in yeast, on the controls of cell cycle division and of DNA repair mechanisms; and the continued developments in understanding the superfamily of P450 genes. Some population-based studies of familial cancer have been reported from the Mormon religious group in the state of Utah (Cannon-Albright et al. 1994) and in Iceland (Tulinius et al. 1994). In the United States, regulatory and quality control of cytogenetic and molecular genetics laboratories is finally underway with the combined forces of the federal government and a professional society known as the American College of Clinical Genetics.

Lagging behind on the 1986 list of recommendations are the coordinated depositing of DNA, the exploration of ecogenetic determinants of cancer beside P450, and the maximal use of twin studies. Draft guidelines on the use of archival pathology material seem very restrictive (ELSI 1994).

Progress is definitely more uneven in the area of administrative and educational measures. Molecular geneticists are certainly populating cancer research institutes worldwide. In contrast, clinicians with the appropriate genetics knowledge and perspective cannot be identified at most such institutions and, hence, are not becoming part of cooperative clinical trials. As some celebrities become identified as members of cancer families, there is a modicum of attention paid to ovarian cancer (Gilda Radner), pancreatic cancer (Jimmy Carter), and breast cancer (film stars and wives of government leaders). A major stimulus is the assignment of some research funds from the NIH's National Center for Human Genome Research dedicated to ethical, legal, and social issues, including ways to introduce predictive gene testing for cancer predisposition into the general population.

In North America, the existence of a uniquely prepared genetics professional, known as the genetics counselor or genetics associate, has been a tool for getting cancer genetics into clinical settings. Such individuals, currently numbering about 1000, are trained to a Master's level in the science and art of medical genetics and counseling and, to date, have been employed largely for prenatal and pediatric genetics. One genetics counselor recently released a monograph on counseling for cancer (Schneider 1994), and an alliance of oncology nurses, counselors, and physicians is being launched to improve communication among the diverse health care professionals interested in the field. In 1995, three short courses on cancer genetics for clinicians will be held in the US alone.

In my opinion, the least progress of all has been made in the actual clinical identification of individuals and families that are likely to benefit from genetics evaluation. In my metropolitan area of 1.5 million persons in Pittsburgh, PA, I see perhaps one consultation a week because of a family history of cancer, in addition to a monthly interdisciplinary clinic for endocrine neoplasia syndromes and a biweekly clinic for the neurofibromatoses. A new initiative on family cancer has been launched by the Swiss Cancer League (Müller 1985; Weber et al. 1995) and endorsed as a global priority by the federation of voluntary national cancer societies, called the UICC. Workshops, symposia, publications and, in the US, legal actions, must be relied upon to set a new standard of care if clinicians or public health officials are going to achieve cancer control through genetics.

Acknowledgement. Supported in part by American Cancer Society grant EDT-80.

References

Albano WA, Lynch HT, Recabaren JA, Organ CH, Maillard JA, Black LE, Follett KL, Lynch J (1981) Family cancer in an oncology clinic. Cancer 47: 2113-2118
Boveri T (1914) Zur frage der entstehung maligner tumoren. Gustav Fischer, Jena
Cannon-Albright LA, Thomas A, Goldgar DE, Gholami K, Rowe K, Jacobsen M, McWhorter WP, Skolnick MH (1994) Familiality of cancer in Utah. Cancer Res 54:2378-2385
Cavenee WK, White RL (1995) The genetic basis of cancer. Sci Am 272(3): 72-79
Chaganti RSK, German J (1985) (eds) Genetics in clinical oncology. Oxford University Press, New York
ELSI (Ethical, Logical, and Social Implications Office) (1994) Informed consent for genetic research on stored tissue samples--draft. National Center for Human Genome Research, 22
Hoskins KF, Stopfer JE, Calzone KA, Merajver SD, Rebbeck TR, Garber JE, Weber B (1995) Assessment and counseling for women with a family history of breast cancer. JAMA 273: 577-585
Li FP, Fraumeni Jr JF, Mulvihill JJ, Blattner WA, Dreyfus MG, Tucker MA, Miller RW (1988) A cancer family syndrome in twenty-four kindreds. Cancer Res 48: 5358-5362
Love RR, Evans AM, Josten DM (1985) The accuracy of patient reports of a family history of cancer. J Chron Dis 38: 289-293
Lynch HT (1994) (ed) Hereditary cancers. Hem/Onc Ann 2: 102-182
Lynch PM, Lynch HT (1985) (eds) Colon cancer genetics. Van Nostrand & Reinhold, New York
Malkin D (1994) Germline p53 mutations and heritable cancer. Ann Rev Genet 28: 443-465
McKusick VA (1994) (ed) Mendelian inheritance in man. 11th Ed. Johns Hopkins University Press, Baltimore
Mitelman F (1991) Catalog of chromosomes aberrations in cancer. John Wiley & Sons, New York
Müller HJ (1985) Familial cancer in Basel: some aspects. In: Müller HJ, Weber W (eds) Familial Cancer. Karger, Basel, 1-5
Müller HJ, Weber W (1985) (eds) Familial cancer. Karger, Basel
Mulvihill JJ (1985) Clinical ecogenetics: Cancer in families. N Engl J Med 312: 1569-1570
Mulvihill JJ (1989) Prospects for cancer control and prevention through genetics. Clin Genet 36: 313-319

Mulvihill JJ (1993) Childhood cancer, the environment and heredity. In: Pizzo PA, Poplack DG (eds) Principles and practice of pediatric oncology. Second Ed. JB Lippincott, Philadelphia, 11-27

Mulvihill JJ (1994) Clinical ecogenetics of human cancer. Hem/Onc Ann 2: 157-161

Mulvihill JJ (1995) McKusick's Mendelian inheritance in man for oncology. Johns Hopkins University Press, Baltimore, in press

Mulvihill JJ, Miller RW, Fraumeni JF Jr (1977) (eds) Genetics of human cancer. Raven Press, New York

Parry DM, Berg K, Mulvihill JJ, Carter CL, Miller RW (1987a) Strategies for controlling cancer through genetics: report of a workshop. Am J Hum Genet 41: 63-69

Parry DM, Mulvihill JJ, Miller RW, Berg K, Carter CL (1987b) Strategies for controlling cancer through genetics. Cancer Res 47: 6814-6817

Philippe P (1989) (ed) Les families à cancer. Les Presses de l'Université de Montréal, Montréal

Rabbitts TH (1994) Chromosomal translocations in human cancer. Nature 372: 143-149

Schneider KA (1994) Counseling about cancer: strategies for genetic counselors. Graphic Illusions, Dennisport, Massachusetts

Tulinius H, Olafsdottir GH, Sigvaldason H, Tryggvadottir L, Bjarndottir K (1994) Neoplastic diseases in families of breast cancer patients. J Med Genet 31: 618-621

Weber W, Narod S, Mulvihill JJ (1995) (eds) Familial cancer. CRC Press, Boca Raton, in press

Human Genome Research and Its Possible Applications to the Control of Genetic Disorders

V.S. BARANOV

Introduction

Tremendous progress in human genome research, especially in the mapping of genes responsible for a number of inherited disorders, has made a substantial impact on early diagnosis and prevention of many genetic disease. This report briefly describes some fundamental and practical areas of human molecular genetics dealing with this problem.

The issues are presented in the form of general answers to four main questions that stem from basic achievements in molecular biology of human genome studies and their practical applications: 1) Where are we now in human genome studies? 2) How might genetic disorders be prevented? 3) What social and ethical problems should be considered in human genome studies and medical genetics counselling? 4) Is human gene therapy a real way out?

General considerations based on the worldwide level of genetic knowledge are supplemented with relevant data from human genome studies and their application in Russia.

Where we are now in Human Genome Studies?

According to a major task of the International Human Genome Project, all primary sequences of human genome DNA, comprising about 3,000 million nucleotides, will be completed by the year 2006. By the end of 1994, the world's collection of nucleotide sequence data released by GenBank consists of only about 200 million base pairs (bp) of the human genome. Meanwhile the elaboration of sequence automation techniques, the development of sophisticated computational tools and drastic technology innovations have rapidly accelerated database growth.

Curiously, there is still controversy about the number of genes contained in the human genome. The quickly expanding knowledge of gene structure, regulation and functions pose many questions as to how to interpret the term, "counting gene." For the sake of simplicity, in the human genome programme

K. Berg, V. Boulyjenkov, Y. Christen (Eds.)
Genetic Approaches to Noncommunicable Diseases

the "counting gene" is considered to be one transcribed unit of DNA that might be translated to one or two functionally linked amino acid sequences (Field 1994). Until recently, the most commonly accepted estimate was that there might be about 100,000 genes altogether in the human genome. However, many people argue that there are probably far fewer.

Reassociation experiments with mRNA populations suggest the existence of 30,000–40,000 genes. However, difficulties in obtaining full representation of low abundance RNAs from all cell types make it likely that this number is an underestimate.

It is well known that all housekeeping genes and at least 40% of tissue-restricted genes are associated with CpG islands. A recently developed experimental approach makes possible the direct separation and counting of all CpG islands (Antequera and Bird 1993). There are about 45,000 CpG islands per haploid genome; thus the total number of genes in humans is estimated at about 80,000. Very similar numbers (60–70,000) were reported recently with the assistance of expression sequence tag techniques (Chen et al. 1994). There are about 4,500 structural genes mapped so far; thus there is still a long way to proceed to the end of the 1.5-meter-long DNA thread of each human cell! Actually this journey does not seem to be so long and desperate (see below).

Recent advances in gene mapping have been mostly confined to dramatic progress in human genetic linkage mapping. After DNA polymorphic sites were identified as suitable molecular markers for gene targeting, it was postulated that only about 150 polymorphic sites, evenly distributed through the human genome, are sufficient for the mapping of any genetic trait. In 1987, the Collaborative Research Group from MIT (USA) published the first linkage map of the human genome, consisting of some 400 polymorphic DNA markers unevenly distributed throughout 23 human chromosomes. Since that time, highly polymorphic minisatellites have become the favoured molecular markers in human genome mapping. The second-generation linkage map encompassing the whole human genome has been constructed based on segregation analysis of 814 highly polymorphic DNA loci containing short CA repeats (Weissenbach et al. 1992). The average resolution of this map (distance between adjacent markers) was estimated as 5 cM - that is, four time higher than is recommended for linkage mapping of any genetic trait (Botstein et al. 1980). Somewhat later in 1994, Weissenbach and collegues at Genethon in Paris presented a new genetic linkage map containing a total of 2,066 short tandem repeats (STS), with an average spacing of just 2.9cM (Gyapay et al. 1994). This map fulfilled one of the main goals of the Human Genome Project, namely, to have a 2- to 5-cM linkage map by 1995! The high informativeness of STR markers (60% of which show a heterozygosity of over 0.7) and the availability of abundant, multikindred families DNA collections create a unique opportunity to map any Mendelian trait and, in particular, any monogenic human disease. The way is now open for comprehensive linkage mapping of multifactorial diseases, that is of genes that predispose to the common multifactorial disorders such as diabetes, cancer, hypertension, heart diseases, etc. It should be mentioned that linkage to a 5-20cM region per se can hardly help much in the identification of the disease-associated mutations (Todd 1992). Such identification requires finer mapping by exploiting linkage disequilibrium between marker polymorphisms and the disease-causing mutations.

These needs, as well as the crucial problem of all human gene identification, might be efficiently approached by new, sophisticated molecular technology for identifying the genes expressed in particular tissue cells by short stretches of their cDNA, called "expressed sequence tags" (EST). Isolated fragments of cDNA, their sequencing and subsequent chromosomal and physical mapping on YAC contig collection of overlapping DNA clones provide the most straightforward approach for mapping and fishing out all human genes (Chen et al. 1994; Marshall 1994). Batallion of 80 automatically operated sequence machines and collection of 300 cDNA libraries from different tissues provided a dramatic outburst of EST identification and mapping. By October 14, 1994, the Institute of Human Genome Research (TIGR) and Human Genome Science (HGS) announced the identification of DNA fragments of 35,000 genes and expected to increase EST numbers by 200, 000 by the end of 1994 (Marshall 1994).

Placing these unique DNA sequences on a physical map with resolution of about 100,000 base pairs constitutes one of the major current tasks of the Human Genome Organization (HUGO) Project, and might be considered as an indispensable prerequisite for the quick and efficient identification of all expressing human genes.

Thus both strategic approaches in human genome studies- 1) the mapping of genetic (Mendelian) traits and the identification of expressing genes with the assistance of STR, STS or EST and 2) direct sequences of all human chromosome DNA are running quite efficiently and substantiate the claims to accomplish the major goals of the International Human Genome Project by 2006.

One should remember, however, that even after this extensive information about the human genome is stored, many problems will remain to be solved. Some that are facing scientists right now involve conspicuous genome diversity, which has already received special attention from HUGO and WHO and resulted in the establishment of the Human Genome Diversity Project (Ginter 1992). Other problems concern the genome and its participation in cell functions, regulations, evolution, etc. Comparative analysis of several totally sequenced genomes of different species-the best current candidates are laboratory mouse (Dietrich et al. 1994), bakers yeasts, and the Nematoda (*C. elegans*; Bargmann 1992) could help to understand the major laws of genome organization and philogeny and extrapolate them to the evolution of life. Maybe someday it will result in the creation of a General Periodical System of All Feasible Genomes, with each species settled into its own separate box as each chemical element is now in Mendeleev's Periodic Table!

How Might Genetic Disorders Be Prevented?

Some of the major results of molecular studies of the human genome have been the mapping, identification and sequencing of a number of genes whose mutations are responsible for many of most common monogenic diseases, such as cystic fibrosis (CF), Duchenne muscular dystrophy (DMD), haemophilia (H), phenylketonuria (PKU) and others. Of approximately 4,500 genes already mapped in human, at least 600 are known to be involved in different inherited diseases. About 20 disease genes are identified in each of 22 autosomes, and over 100 of them are mapped to the X chromosome.

Efficient direct and indirect (confined to molecular marker) methods have been elaborated for precise molecular diagnosis of many monogenic diseases. Identification of a number of major mutations or particular types of mutations by means of the relatively simple, reliable and inexpensive polymerase chain reaction (PCR) method, or by means of some of its numerous variants, opened the way for population studies of inherited diseases. In addition to efficient biochemical tests of newborns for PKU, hypothyrioidism, and galactosemia, molecular screening programmes for CF, DMD, B-thalassemia, sickle-cell anaemia, and Gaucher disease have been launched in many developed countries. DNA analysis and protein analysis may provide data of comparable quality, as in the case of sickle cell anaemia and α.1-antytripsin deficiency. In other cases, protein or enzyme analysis combined with DNA analysis is more effective than either one alone, as in the case of Tay-Sachs disease. In other instances, DNA analysis is the only method available at present. Such as for carrier detection for CF, Robotics-assisted procedure for large-scale (20) common CF mutations screening is suggested (DeMarchi et al. 1994).

Molecular analysis has been found to be very productive not only for presymptomatic diagnosis of many inherited disorders, but also, for asymptomatic carrier detection in relevant high-risk families.

Efficient schemes of molecular analysis are widely applied for prenatal diagnosis of many still incurable and lethal postnatally inherited disorders. Here in Russia prenatal diagnosis is recommended to families who are at-risk for least 20 different monogenic disorders, and this list might be quickly expanded (Baranov 1993).

Thus population and high-risk families screening for some major mutations of common genetic disorders, in conjunction with efficient prenatal (including preimplantation) molecular diagnosis and supplemented with highly professional medical genetic counselling, should be taken as very efficient means for the control and prevention of genetic disorders.

The current level of human genome knowledge, as discussed above, opens the door for comprehensive mapping and identification of genes that predispose to common human multifactorial disorders (diabetes, cancer, hypertension, heart disease), some of which have already been identified (breast cancer, diabetes, adenomatosis polyposis, hypercholesterinemia and many others).

Considering that so many major mutations of monogenic disorders, as well as pathological alleles of the genes predisposing to mulifactorial diseases, might be identified relatively easily after birth, one should think about the pros and cons of a "genetic certificate" for each newborn. The validity of such a certificate and its benefits for the health and welfare of the individual, close relatives and physicians (family doctor) remain questionable and deserves serious consideration.

Some Social and Ethical Problems in Human Genome Studies

The Human Genome Organisation Project has been considered to be rather costly since its onset in 1987, when the price for one DNA step (nucleotide base pair) was estimated at about one American dollar. Subsequent technological improvements, automation and technical innovations reduced this price by half by 1994. Nonetheless the clear-cut benefits to all mankind of the knowledge of the human genome have stimulated an exponential growth of information, with more than half of the human genome sequences being added in the last two years. Two recent events have caused a substantial mess in this otherwise calm field of international collaboration. Both concern the applications by some American scientists involved in HUGO to patent the results of their genome studies. One of the scientists, Mark Skolnick from the University of Utah and owner of a commercial firm, Myriad Genetics, declared his decision to patent the gene BRCA-1, which is responsible for breast cancer susceptibility. The other, Craig Venter at The Institute for Genome Research (TIGR), suggested the patenting of gene fragments (ESTs; see above). Both of these applications have been thoroughly discussed in the scientific literature (Marshall 1994; Dickson 1994; Buttler and Gershon 1994), with many leading officials of the HUGO Project (T. Caskey, F. Collins) opposing the applications.

The other social problem that jeoparadises the application of human genome findings concerns the cost of molecular tests, both for affected and healthy people. Thousands of families throughout the world are at-risk for already well-known disease genes, costs of testing for which fluctuate between 200 and 1,000 according to some American companies (Novak 1994). Therefore, the major bulk of the world population, especially in developing countries, will hardly have access to these already ripe fruits of the HUGO tree!

Another serious problem awaiting an urgent solution concerns the dissemination of genetic knowledge. This issue was thoroughly reviewed recently in a number of papers (Rennie 1994) and discussed at this WHO meeting (Knoppers 1994). Briefly, the results of a wide spread screening programme for sickle cell anaemia in the USA were sometimes misused to discriminate against healthy carriers of the trait. Much better public education about the meaning of the tests, greatly helped in a screening programme for Tay-Sachs disease. Now the positive experience of the Tay-Sachs programme is being used in a CF screening programme.

Meanwhile, many ethical and social problems in gene screening programmes remain. The main one is genetic privacy. Who should be informed of the results of genetic testing? In a 1992 poll, a majority of Americans (57%) said the privacy of test results should not be absolute. The problem stems from the fact that your genes in a strict sense are not exclusively yours; you share half of them with your parents, siblings and children. Thus if you discover a mutated allele in your genome, you may have an ethical, if not a legal, obligation to inform your relatives (Rennie 1994). Genetic privacy as a prerequisite of personal autonomy therefore comes into definite conflict with the need to share this information with a spouse or fiance, other family members, family doctors and even with insurers or employer. The situation seems even more desperate if one considers the feasible ethical and social consequences of a Genetic Certificate, mentioned in the previous section. What is preferable – to live as an ostrich with its head plunged in the sand or to know from childhood the strong and weak points of your genome and thus be able to create a robust and rational life style? If not provided to the child, should the parents or family doctor be provided with a genetic certificate at the child's birth, to ensure that the child is provided adequate growing and education facilities?

Thus with the rapid growth of our knowledge of the human genome, and especially of the parts responsible for numerous genetic disorders of monogenic and polygenic origins, the community is faced with more and more social and ethical problems that should not be postponed and need urgent solutions, if only to avoid misunderstandings comparable to earlier misinterpretations of eugenic principles initially presented by Francies Galton. More efficient genetic education should be provided not only for physicians but also for the lay public.

Is Human Gene Therapy a Real Way Out?

Dramatic progress in the HUGO Project, conspicuous advances in the understanding of molecular biology of human diseases and the development of efficient gene transfer techniques have resulted in practical approaches to human gene therapy (Miller 1992). Since the first human gene therapy trial in September 1990, dealing with the treatment of adenosine-deaminase deficiency by ex vitro transfer of ADA gene into a patient's T- cells, the number of ongoing and already approved clinical trials for genetic disorders is now approaching 100 (Culver 1994). The results of these trials, as well as of other scientific aspects in this quickly expanded field, are published in the journals, *Gene Therapy* and *Human Gene Therapy*. Among the ongoing and approved candidates for gene therapy are almost all of the common monogenic diseases, some cancers, multifactorial disorders and even some virus infections (HIV, hepatitis B). All of these trials are carried out on postnatal patients, with all human gene transfer protocols adopted by corresponding ethical and advisory committees. All genetic constructions used for gene therapy are subjected to relevant genetic trials on suitable animal models.

In addition to the already reported ADA treatment by transfer of a corresponding gene, very encouraging results have been obtained with treatment of hypercholesterinemia by transfer of a LDL receptor gene into hepatocytes and by transfer of a tumor necrosis factor (TNF) gene to tumor-infiltrating lymphocytes (Miller 1992). Regardless of gene function and the tissue of its expression, each genetic diseases needs its unique protocol of gene therapy. Conspicuous recent advances in this field are confined to the first attempts to use the gene therapy approach for the treatment of CF, the most common monogenic disease in Caucasians. Several clinical trials of CF gene therapy have already been undertaken in the USA, UK and France. The results, though not unanimously positive, should nevertheless be considered quite encouraging, and indicate feasible ways to improve gene delivery and expression in CF patients (Crystal et al. 1994; Alton and Geddes 1994)

The gene therapy approach looks very plausible for correction of gene disorders in haemophilia B, phenylketonuria and DMD, although many technical problems should be resolved before clinical trails are undertaken.

Although it is a decisive step towards ultimate control of genetic disorders, gene therapy by itself can hardly be considered a real way out of major problems facing medical genetics at present. Moreover, this utilisation of DNA as drugs should result in a very quick expansion of affected genes in human populations, as it is still confined and will be confied for the indefinite future to somatic cells but not to germ cells or to early human embryos. The quick insemination of the human population with formerly lethal mutations of CF, PKU and may other genetic disorders has already started.

More sophisticated, reliable and cheap screening programmes, and efficient prenatal diagnosis combined with adequate medical genetic counselling, should be applied more widely to resist the flooding of the human genome with disease genes.

References

Alton E, Geddes D (1994) A mixed message for cystic fibrosis gene therapy. Nature Genet 8: 8–9

Antequera F, Bird A (1993) Number of CpG islands and genes in human and mouse. Proc Natl Acad Sci USA 90: 11995–11999

Baranov VS (1993) Molecular diagnosis of some common genetic diseases in Russia and the former USSR: present and future. J Med Genet 30: 141–146

Bargmann CI (1992) cDNA sequencing: a report from the worm front. Nature Genet 1: 79–80

Botstein D, White RL, Skolnick MH, Davis RW (1980) Construction of the genetic linkage map in man using restriction fragment length polymorphisms. Am J Human Genet 32: 314–331

Buttler D, Gershon D (1994) Breast cancer discovery sparks new debate on patenting human genes. Nature 371: 221–222

Chen E, d'Urso M, Schlessinger D (1994) Functional mapping of human genome by cDNA localisation versus sequencing. BioEssay 16: 693–698

Crystal RG, McElvaney NG, Rosenfeld M, Chu CS, Mastrangeli A, May JG, Brody SL, Jaffe HA, Eissa NT, Panel C (1994) Administration of an adenovirus containing the human CFTR cDNA to the respiratory tract of individuals with cystic fibrosis. Nature Genet 8: 42–51
Culver KW (1994) Gene Terapy Mary. Ann Lieberth Inc Publ
DeMarchi JM, Richards CS, Fenwick RG, Pace R, Beaudet AL (1994) A robotics-assisted procedure for large scale cystic fibrosis mutation analysis. Mutat Res 4: 281–290
Dietrich WF, Millere JC, Steen RG, Merchant M, Damron D, Nahf R, Gross A, Joyce DC, Wessel M, Dredge RD, Marquis A, Stein LD, Goodman N, Page DC, Lander ES (1994) A genetic map of the mouse with 4,006 simple sequence length polymorphism. Nature Genet 7: 220–246
Dickson D (1994) Ownership and the human genome. Nature 371: 363–366
Field J (1994) How many genes in the human genome? Nature Genet 7: 345–348
Ginter EK (1992) The human genome diversity project. WHO Report 9: 1–3
Gyapay G, Morissette J, Vignal A (1994) The 1993–94 Genethon human genetic linkage map. Nature Genet 7: 246–339
Knoppers BM (1994) Towards an ethics of "compleity" for "common" disease. Genetic approaches of common diseases (Berg K, Boulyjenkov V, Christen Y eds). Berlin, Heidelberg, Springer Verlag
Marshall E (1994) A showdown over gene fragments. Science 266: 208–210
Miller AD (1992) Human gene therapy comes of age. Nature 357: 455–460
Novak R (1994) Genetic testing set for takeoff. Science 265: 464–467
Rennie J (1994) Grading the gene tests. Sci Am 270 (6):88–97
Todd JA (1992) La carte des microsatellites est arrive! Human Mol Genet 1: 663–666
Weissenbach J, Gyapay G, Dib C, Vignal A, Morissette J, Millassean P, Vaysseix G, Lathrop M (1992) A second-generation linkage map of the human genome. Nature 359: 794–801

Improvement of Adenoviral Vectors for Human Gene Therapy

E. VIGNE, J.-F. DEDIEU, C. ORSINI, M. LATTA, B. KLONJKOWSKI, E. PROST, M.M. LAKICH, E.J. KREMER, P. DENÈFLE, M. PERRICAUDET, and P. YEH

Summary

The transfer of therapeutic genes into somatic cells will have an application in the treatment of a variety of acquired as well as hereditary diseases. Retroviral vectors are being evaluated in many gene transfer clinical protocols. However, these vectors suffer inherently from their requirement for dividing cells for efficient transduction. The availability of a clinical grade vector that could efficiently transduce highly differentiated resting cells (e.g., quiescent lymphocytes or normal epithelial cells) will broaden the number of future gene therapy indications. Other modes of gene transfer, including various means of transfection and other viral vectors, are also being intensively evaluated. Some of our strategies to develop safer and more efficient adenoviral vectors (adenovectors) for clinical application will be reviewed. For example, in the course of designing the third generation of adenovectors (i.e., deleted for 2 functions required for viral propagation), several systems were compared with respect to their ability to transcomplement the E1 and E4 adenoviral functions. In particular, a 293 $E4^+$ cell line was constructed that allowed plaque purification and propagation of highly defective adenovectors expressing the β-galactosidase-encoding gene (lacZ) in different E1 and E4 deletion backgrounds. Also, to increase the safety profile of adenovectors in humans, especially regarding their possible rescue following gene transfer, a non-human (canine) adenovector containing the lacZ gene has been constructed that demonstrates efficient gene transfer into human cells.

Introduction

Gene therapy can be described as the introduction of genetic material (i.e., DNA or RNA) of therapeutic value into a target cell population, either to enhance a cellular function or to confer a new property to the target cells. The excitement of the scientific community for this approach is mirrored by increasing research and development. Novel therapeutic strategies directed towards a wide range of

K. Berg, V. Boulyjenkov, Y. Christen (Eds.)
Genetic Approaches to Noncommunicable Diseases

acquired as well as genetic diseases will hopefully follow. Gene therapy is also likely to broaden its field of investigation as genes involved in human diseases are continuously being identified and cloned. However, it cannot be overstated that at present, gene therapy is very much in its infancy.

Because viruses ("bad news wrapped in proteins") have evolved very efficient means to enter their target cells and deliver their own genetic material, many approaches to the gene transfer challenge make use of defective viruses. Since the ultimate target of gene delivery is a living organism, viral vectors deserve exhaustive attention so that they can be turned into "good news." Since the first approved clinical trial 5 years ago, which relied on the use of a recombinant retrovirus to introduce the human adenosine deaminase gene into the T-cells of 2 children (Culver et al. 1991), the number of gene therapy proposals has increased exponentially (for a review, see Freeman et al. 1993). Although no morbidity has ever been reported from virus-based gene transfer during clinical trials, safety issues have not yet been resolved. In this review, the specific advantages and drawbacks associated with the viral vectors currently in use, or approved for clinical evaluation, will be discussed. The main features of herpes-derived vectors have therefore been excluded.

Viral Vectors for Long-Term Expression

Retroviral Vectors

Over the past decade, retroviral-mediated gene transfer (RMGT) has emerged as a very efficient way to randomly insert new genetic material into dividing somatic cells. RMGT results in the permanent *de novo* addition of genetic information that will be subsequently transmitted to the cellular progeny. Several features merit discussion. First, infectious, although replication-defective, recombinant retroviral stocks are generated in a packaging cell line encoding all the viral proteins required for virus production. To reduce the emergence of replication competent retroviruses (RCR) during viral stock preparation, the trans-complementing viral genes of the packaging cell line have been dispatched in several locations in the cellular genome. However, emergence of RCR may still occur as a result of several independent homologous recombination events between the many vector backbones and residual viral sequences flanking the transcomplementing genes of the packaging cell lines (Otto et al. 1994). Second, current technology allows titers of 10^7–10^8 colony-forming units (CFU)/ml of recombinant retroviruses (Kotani et al. 1994). Unfortunately, these dilute viral suspensions are not satisfactory to achieve efficient in vivo gene transfer. Although the direct delivery of recombinant retroviruses has been reported in dogs following partial hepatectomy (Ferry et al. 1991), RGMT is obviously more efficient during consecutive rounds of viral transduction ex vivo. Third, the cloning capacity of these viral vectors is limited to less than 6 kb, which may be too small for a number of candidate genes.

A *theoretical* concern inevitably associated with retroviral vectors is that insertional mutagenesis (of an oncogene or a tumor suppressor gene, for example) could occur following gene transfer. However, tumor formation has never been associated with retroviral vectors that are *not* contaminated with RCR (Anderson 1994), probably because multiple independent integration steps are required to induce an uncontrolled proliferative phenotype in humans (Weinberg 1989). In fact, the development of a retrovirus-induced T-cell lymphoma has been reported, but only in a limited number of severely immunosuppressed monkeys, and only in those animals that were exposed to *very high doses of RCR* (Payne et al. 1982). A more limiting drawback of retroviruses is that they depend on host cell division for integration and long-term expression of the transgene. This therefore limits their utilization to circumstances in which cell division occurs in a high proportion of cells (Ferry et al. 1991). This requirement is particularly important because many putative targets for human gene therapy are actually post-mitotic cells.

Adeno-associated Viral Vectors

Five human adeno-associated virus (AAV) serotypes have been described so far. They are widely prevalent in man (more than 90% of U.S. adults are seropositive) but they have not been associated with any clinical symptoms (for a review, see Leonard and Berns 1994). An interesting feature is that all human cells so far tested can be successfully transduced by AAVs in vitro. Contrary to retroviruses, AAVs can integrate their DNA into non-dividing cells. However, AAV-mediated transduction has been reported to be several orders of magnitude more efficient in dividing cells as compared to quiescent cells, at least for primary human fibroblasts (Russell et al. 1994). Also, DNA-damaging agents have been reported to increase gene transfer up to 750-fold in nondividing, and to a lesser extent, in dividing fibroblasts (Alexander et al. 1994). Taken together, these results strongly suggest that quiescent cells are indeed much less susceptible than proliferative cells to AAV-mediated transduction. Human AAVs also have intrinsic limitations due to their requirement for helper virus (adenoviruses or herpes viruses) key functions during viral stock preparation (i.e., viral propagation). Adenoviral helper functions are the best characterized and include the E1a proteins, the E1b 55 kD protein, the gene product of E4 open reading frame 6 (ORF6, see below), the E2a 72 kD DNA binding-protein, and the virus-associated (*VA*) RNAs. Without such helper functions human AAVs do not propagate, but instead integrate into the host genome where they remain as proviruses (for a review, see Leonard and Berns 1994). The only sequences required for AAV integration lie within their terminal 145 nt inverted terminal sequences, so that their maximal cloning capacity is less than 5 kb. Integration of wild-type AAV into the host genome is targeted by the AAV *rep* proteins to location 19q13.3 (Giraud et al. 1994; Samulski et al. 1991; Weitzman et al. 1994). On the other hand, recombinant AAV vectors are *rep*$^-$ and integrate randomly. Therefore, contrary to retroviral vectors that incorporate their own integrase during viral propagation, AAVs apparently do not incorporate the *rep* proteins within their capsid. Most importantly, AAV and retroviral vectors do not carry residual viral genes in their backbone, and consequently do not express any viral antigens following gene transfer. This finding and the fact that they integrate their genetic information in the cellular genome allow for a sustained expression of the transgene that they carry (Flotte et al. 1993; Walsh et al. 1994). However, like the retroviruses, a limitation of AAV-mediated gene transfer is the extremely low titers currently obtained for viral stock preparation (for a review, see Kotin 1994).

Adenoviral Vectors for Gene Transfer

First-generation Human Adenoviral Vectors

There are many reasons to use recombinant adenoviruses (rAd or adenovectors) for human gene therapy. First, adenoviruses have been used as live enteric vaccines for many years with an excellent safety profile. Furthermore, attempts to establish adenovirus as a causative agent in human cancer have been uniformly negative. Also, type 2 (Ad2) or type 5 (Ad5) adenoviruses (subgroup C) are extremely common in humans and cause only benign respiratory symptoms in immunocompetent individuals (Horwitz 1990). Second, most adenovirus serotypes are believed to use a ubiquitous cell integrin (the vitronectin receptor $\alpha_V\beta3$ or $\alpha_V\beta5$) to enter the target cells (Mathias et al. 1994; Nemerow et al. 1993). Interestingly, cell entry of Ad2 (and presumably Ad5) adenoviruses may not rely solely on this particular cellular integrin (Bai et al. 1994); the possibility that these adenoviruses, from which current vectors are derived, interact with multiple receptors (possibly related integrins) for cell internalization would actually explain why adenovirus-mediated gene transfer (AMGT) can be achieved in many cells of many organs, and in many animal species (for a review, see Kozarsky and Wilson 1993; Trapnell and Gorziglia 1994). Third, the adenoviral genome is 36 kb in length (Fig. 1), so that large DNA inserts that could not be introduced into AAV or retroviral vectors can be expressed within an adenoviral construct (Ragot et al. 1993), possibly from a large tissue-specific promoter. Finally, transgene expression from the vector is achieved from dsDNA without integration into the host cell genome, alleviating concerns about insertional mutagenesis events. Furthermore, this also explains why AMGT and expression is achieved with no requirement for cell proliferation.

First-generation adenovectors are replication-defective because they lack a 3 kb fragment from the E1 region required for viral propagation (Fig. 1). The technology to produce viral stocks of otherwise non-replicative adenoviral vectors therefore relies on an artificial packaging cell line to provide the missing E1a and E1b regulatory proteins. This transcomplementation is usually achieved using the 293 cell line, a human embryonic kidney cell line in which Ad5 adenoviral sequences, expressing the E1 genes together with E1 flanking sequences, have been integrated in the cellular genome (Graham et al. 1977).

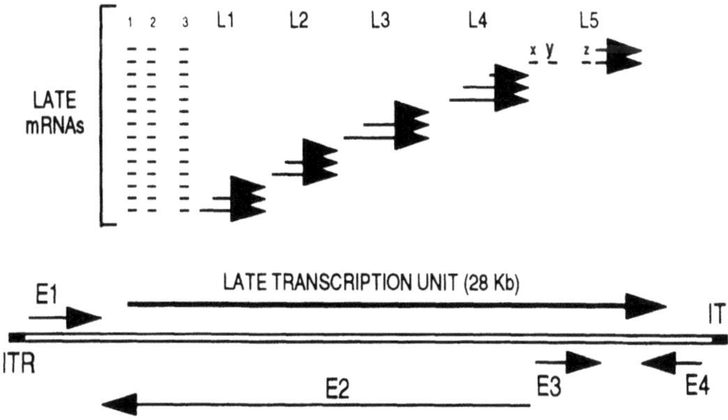

Fig. 1. Schematic diagram of the Ad5 genome and its major transcription units. The Ad5 genome is a linear 36 kb dsDNA molecule. The main transcription units include four distinct early regions (*E1-E4*) and a late transcription unit (LTU) that is alternatively polyadenylated and spliced to generate five (*L1-L5*) distinct families of late mRNAs (*upper panel*). Late proteins derived from the L1-L5 mRNAs are mostly structural virion proteins. The switch from early to late gene expression takes place approximately seven hours after infection. E1 proteins activate most other viral transcription units either directly or indirectly. E2 proteins include the 72 kD DBP, the viral DNA polymerase and the terminal binding protein which are all directly involved in viral DNA replication. E3 proteins modulate the host immune response towards virally infected cells. E4 proteins are involved in the control of viral transcription, viral DNA replication, and viral late protein synthesis. 1, 2 and 3 refer to the components of the tripartite leader present in all late mRNAs; x,y, and z refer to the components of the auxiliary leader present in some of the fiber-encoding mRNAs (L5). The plx- and IVa2-encoding genes and the *VA* RNAs have been omitted for clarity. The E1 region is deleted in first-generation adenovectors and is usually replaced by the transgene expression cassette

Although this technology is extremely efficient for preparing viral stocks with titers above 10^{12} particle-forming units (PFU)/ml, it is not satisfactory for the preparation of clinical grade vectors because recovery of replication-competent adenoviruses (RCA) is possible during viral stock preparation (Lochmuller et al. 1994). As for RCR, the emergence of RCA apparently occurs during viral stock preparation upon homologous recombination between the episomal adenovector genomes and the cellular E1 flanking sequences in 293 cells (Fig. 2, panels A and B). A possible solution to this problem is to design another E1-transcomplementing cell line with no overlapping sequences between the cellular adenoviral sequences and the vector backbones. We took a different approach by remodeling the vector backbone of first-generation adenovectors (Fig. 2, panel C) so that homologous recombination between the 293 genome and the vectors would generate an $E1^{+}E4^{-}$ replication-incompetent virus (manuscript in preparation). Another solution relies on the additional expression of a second essential adenoviral region within the 293 cells and its concomitant deletion within the adenovectors (see below).

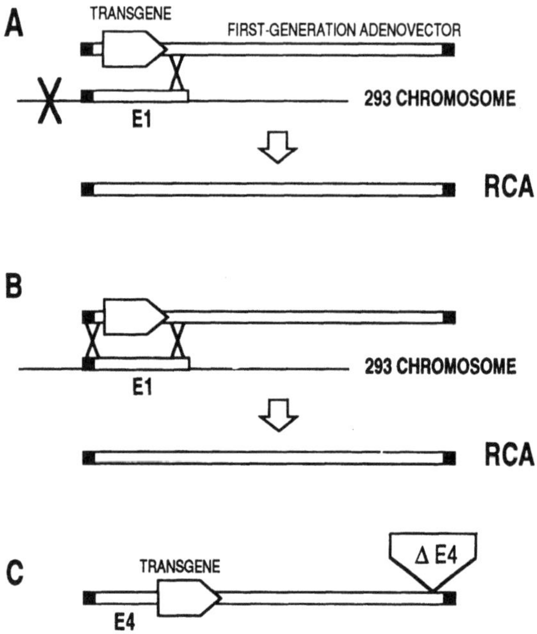

Fig. 2. Generation of RCA during viral stock preparation of first-generation adenovectors in 293 cells. The emergence of $E1^+$ vectors can possibly arise **a** after one homologous recombination event and chromosomal breakage, or **b** two homologous recombination events on either side of the E1 region. **c** Remodeling the vector backbone of first-generation adenovectors by placing the entire E4 region in place of E1 does not regenerate a replication-competent adenovirus (*RCA*) following either mechanism. ΔE4 refers to a deletion inactivating at least both the ORF3 and ORF6 gene products

Further deletions of sequences that are dispensable for viral propagation have been achieved to increase the cloning capacity of first-generation adenovectors. These deletions encompass approximately 2.7 kb of the E3 region (Bett et al. 1993) and 1.4 kb corresponding to the proximal part of the E4 region (Hemstrom et al. 1988) (Fig. 3), extending the cloning capacity of current vectors to approximately 8 kb. In contrast to retroviral or adeno-associated viral vectors, recombinant adenoviruses consequently retain most of the viral genome (approximately 80%, potentially encoding more than 20 proteins). Low level expression of viral genes has been documented following gene transfer in mice (Yang et al. 1994a), cotton rats (Engelhardt et al. 1994b), and human bronchial epithelia of xenografts (Engelhardt et al. 1993). Because the E1A region codes for transactivating transcription factors targeting most adenoviral promoters, this leaky viral expression is probably a consequence of cellular "E1a-like" factors. For example, an E1a-like activity has been documented in undifferentiated F9 embryonal carcinoma cells (Imperiale et al. 1984), which is lost upon cell differentiation (La Thangue and Rigby 1987). Furthermore, activation of the adenoviral E4 promoter has been demonstrated in several organs following intravenous injection in adult mice with an E1-deleted adenovector in which the lacZ reporter gene is expressed from this promoter (N. Hanania, personal communication). In that respect, c-myc and an IL6 (a major inflammatory cytokine)-regulated cellular factor are potential candidates to mediate this transactivating function, as previously proposed (Onclercq et al. 1988; Spergel and Chen-Kiang 1991). Also, the IL6-induced E1a-substituting activity is

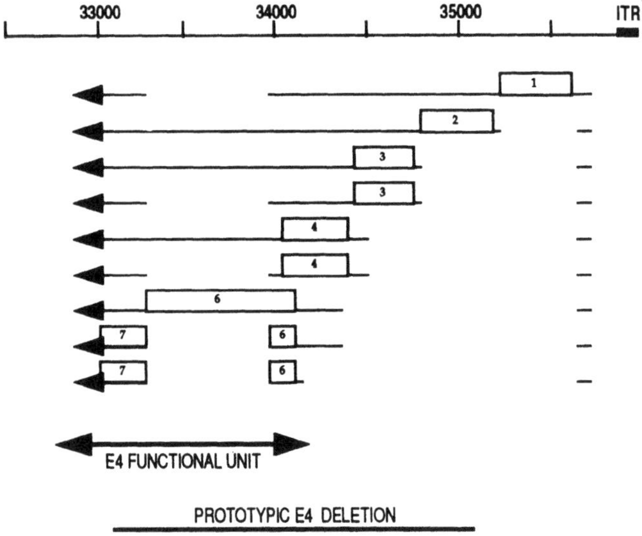

Fig. 3. Genetic organization of the Ad5 E4 region. The main mRNAs encoding E4 ORF1, ORF2, OFR3, ORF4, ORF6 and ORF6/7 gene products are mainly based on the work of Dix and Leppard (1993). The extent of the E4 deletion harboured by dl808 (a prototypic defective E4 deletant missing both the ORF3 and ORF6 proteins) is indicated by a solid bar. The viral sequence encompassing ORF1, ORF2, ORF3, and ORF4 can be deleted without dramatic consequences for viral stock preparation in 293 cells (Hemstrom et al. 1988). Therefore, the distal part of the E4 region including ORF6 and ORF7 constitutes the minimal E4 functional unit that was integrated in the 293E4 cell line (see text). Note the absence of overlapping sequences upstream of E4 ORF6 between the minimal E4 functional unit and the remaining sequences of a defective adenoviral E4 deletant

apparently identical, or related to, the NF-IL6 transcription factor (Spergel et al. 1992). This is significant because elevated levels of IL6 have been reported in one patient following AMGT into the lungs (Crystal et al. 1994). It seems, therefore, conceivable that the local inflammatory response that is commonly associated with AMGT is triggered, and possibly amplified, by leaky viral antigen expression.

Transgene expression following AMGT in immunocompetent adult animals usually does not exceed a few weeks (for a review, see Trapnell and Gorziglia 1994). In fact, a local dose-dependent inflammatory response has been reported by many groups following AMGT (Herz and Gerard 1993; Simon et al. 1993; Yei et al. 1994b), and transgene expression is substantially longer in immunodeprived or immunodeficient hosts (Engelhardt et al. 1994a,b; Yang et al. 1994a). These data suggest that a major component of the rapid extinction of transgene expression involves a CTL-mediated clearance of the infected cells resulting from the MHC-I-restricted presentation of viral antigens (Yang et al. 1994a,b). In that respect, the leaky expression of residual viral genes following in vivo transfer is certainly a major drawback associated with the use of adenovectors (and *a fortiori* with herpes-derived vectors) for gene therapy. The deletion of the

E3 region within the vector backbone is, therefore, certainly undesirable, because this region encodes several proteins whose natural function is likely to counteract independent aspects of the host immune response towards adenovirus-infected cells (for a review, see Wold and Gooding 1991). For example, a 19 kD glycoprotein (gp 19K) prevents CTL-mediated cytolysis by preventing cell surface localization of certain MHC-I alleles (Beier et al. 1994; Flomenberg et al. 1992). The E3 region also encodes proteins that can prevent E1a-induced tumour necrosis factor (TNF) cytolysis (Gooding et al. 1991). Surprisingly, administration of an $E1^-E3^+$ adenovector in cotton rats apparently led to a moderate host response (Zabner et al. 1994). As the E3 promoter is E1a responsive, some E1a-like factors may somehow be involved in the leaky expression of non-deleted viral genes. Further deletion of the viral genome to decrease viral gene expression from an $E1^-E3^+$ adenovector is actually our goal, in order to make AMGT safer and better (see below).

Crippling Adenoviral Expression: Second-generation Human Adenovectors

ts125 is a thermosensitive point mutation in the gene encoding the 72 kD adenoviral DNA binding protein (DBP), which severely impairs the initiation of viral DNA replication at the non-permissive temperature of 39°C (van der Vliet and Sussenbach 1975). A definite improvement in E1-deleted adenovectors has been to further cripple in vivo viral expression by the additional introduction of the ts125 mutation in the vector backbone (Engelhardt et al. 1994a). The technology to produce these doubly defective (second-generation) adenovectors once again relies on 293 cells, as they can be infected at the permissive temperature (32°C) without significant loss of viral productivity (Engelhardt et al. 1994a). Most importantly, transgene expression is significantly prolonged following transfer of this second-generation adenovector into mouse liver, or into the lungs of cotton rats (Engelhardt et al. 1994a,b, Yang et al. 1994b). However, point mutations are subject to reversion and/or they may be partially active at 37°C, as is the case with ts125 (Engelhardt et al. 1994a). These drawbacks, together with the emergence of RCA during viral stock preparation, will be eliminated by further deleting an indispensable adenoviral region from the vector backbone (see below).

Silencing Viral Expression: Third-generation Human Adenovectors

Rationale

Deleting a region required for viral growth implies that the missing adenoviral functions can be provided *in trans* during viral stock preparation, either by a helper virus or in a transcomplementing cell line (see below). The E4 region is a good candidate because it is required for viral propagation, viral late protein synthesis and virus particle assembly (Falgout and Ketner 1987; Halbert et al. 1985; Weinberg and Ketner 1986). Second, the E4 region encodes regulatory functions mostly expressed during the early phase of infection (i.e.,prior to viral DNA replication), so that transcomplementation from a limited number of integrated chromosomal copies is realistic. In fact, W162 is a VERO-derived E4 transcomplementing cell line that allows the viral propagation of highly defective $E1^{+}E4^{-}$ adenoviruses (Weinberg and Ketner 1983). However, E4 transcomplementation is not optimized in this monkey-derived cell line, as judged by a remarkably low PFU/viral particles ratio (E. Vigne, unpublished observation). Finally, the E1 and E4 regions are located on opposite ends of the adenoviral genome (Fig. 1). Multiple recombination events would therefore be required to regenerate a replicative $E1^{+}E4^{+}$RCA.

The E4 region of Ad2 or Ad5 (91.3 to 99.1 m.u.) is expressed throughout a productive infection and is extensively spliced (Virtanen et al. 1984). This indispensable adenoviral region encodes at least six different open reading frames (ORFs),some of which code for proteins of unknown function (see Fig. 3). Importantly, ORF3 and ORF6 encode essential, but redundant, activities, although the ORF6 gene product is more potent than the ORF3-encoded protein for viral propagation (Bridge and Ketner 1989; Huang and Hearing 1989). Both proteins are somehow involved in the overall process of late mRNAs formation, and consequently viral late protein expression, from the 28 kb late transcription unit (LTU; see Fig. 1). For example, these proteins could control alternative polyadenylation and/or splicing of the LTU from which most viral structural proteins are derived. In this respect, they have been reported to modulate different aspects of splicing in short-term transfection assays, and also during the course of a viral infection (Nordqvist and Akusjarvi 1990; Nordqvist et al. 1994; Ohman et al. 1993). Interestingly, these proteins are not homologous and apparently act through different pathways (Bridge and Ketner 1990), a situation somehow reminiscent of the recognition of weak acceptor sites by the Sx1 and Tra proteins involved in the alternative splicing cascade occurring during sexual differentiation in drosophila (Tian and Maniatis 1993; Valcarcel et al. 1993). Most importantly, the redundancy of the ORF3 and ORF6 proteins for E4 function is obviously the molecular basis for the lack of known thermosensitive mutation in E4.

The Helper Virus Approach for E4 Transcomplementation

In the course of designing adenovectors deleted for both E1 and E4, we evaluated several systems for their ability to transcomplement the E4 function. One possibility relies on the use of an E4 transcomplementing minimal helper virus. Briefly, an E4 minimal adenovector was designed that can fully transcomplement the viral function missing in Ad2dl808 (Challberg and Ketner 1981), a prototypic defective E4 deletant missing both the ORF3 and ORF6 gene products (Fig. 3). As shown in Fig. 4, plasmid pE4Gal exhibits all the cis determinants required for viral DNA replication and packaging: the 103 bp

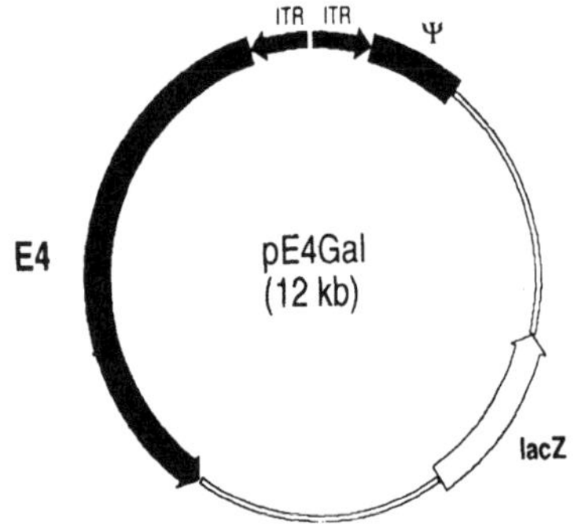

Fig. 4. Design of the helper plasmid/virus pE4Gal. This plasmid encodes all the cis determinants required for viral replication (*ITR*) and packaging (Ψ). It also encodes the whole E4 region expressed from its own promoter. This plasmid can be propagated as a mini-virus in the presence of the dl808 genome (see text)

inverted terminal repeats (ITR) normally located on both ends of the viral genome, and the Ψ sequence, respectively. pE4Gal also expresses the whole Ad5 E4 region from its own promoter (i.e., in an E1a-responsive manner), together with a nuclear-targeted β-galactosidase as a phenotypic marker. When 293 cells are transfected with plasmid pE4Gal, and subsequently infected with Ad2 dl808 ($E1^{+}E4^{-}$), a cytopathic effect (CPE) can be obtained, amplified and plaque purified in 293 cells. This CPE is actually the result of the dual propagation of dl808 and a minivirus derived from plasmid pE4Gal in every infected cell. The two viruses can be separated by consecutive rounds of ultracentrifugation; viral titration of Ad2dl808 on W162 cells also demonstrates that high titers of the dl808 virus can be obtained (10^{11} PFU/ml), and with minimal viral contamination of the $E4^{+}$ minivirus (less than 10^{-6} as judged by X-GAL staining of the infected cells). We concluded that this "one cell/two virus" system is actually very efficient to rescue the E4 function of dl808 during viral propagation in 293 cells. However, this technology is apparently not suitable for the initial step of viral construction of E4-deletants in 293 cells (E. Vigne, unpublished observation).

The Cell Line Approach for E4 Transcomplementation

An alternative exists which relies on an $E1^{+}E4^{+}$ transcomplementing cell line. We anticipated that this would be difficult to achieve due to the very nature of some of the ORFs of the E4 region, which antagonize and/or interact with both E1a and E1b proteins in a timely fashion during the viral cycle. For example, the E1a and E4 ORF6 proteins both interact with E2F, an important cellular component of cell cycle progression (for a review, see La Thangue 1994). The E1b 55 kD protein also associates with roughly half of E4 ORF6 encoded protein, within a nuclear complex somehow controlling the export of a defined subset of (responsive), mostly viral, RNAs from the nucleus to the cytoplasm (Pilder et al. 1986). In addition, the E4 ORF4 encoded protein appears to modulate the phosphorylation of E1a proteins (Muller et al. 1992), apparently leading to the control of viral DNA replication (Bridge et al. 1993).

Because the E1 region is constitutively expressed in 293 cells, we constructed a 293-derived cell line expressing the E4 function from an inducible promoter: expression must be low enough in the uninduced state to avoid toxicity to the host cell, but high enough in the induced state to make enough E4 region gene products to fulfill the many aspects of E4 function in a normal viral cycle. Furthermore, as the ORF6 protein is more potent than the ORF3 protein for E4 function, and because we feared that high level expression of all the E4-encoded proteins may be deleterious for cell viability, a "minimal" E4 functional unit was designed in which the distal part of E4 (potentially encoding for the ORF6 and ORF6/7 gene products; see Fig. 3) is expressed from the glucocorticoid-inducible MMTV promoter. Transfection of this construct in 293 cells allowed the recovery of a stable cell line that can efficiently transcomplement for both the E1 and E4 functions required for viral propagation (submitted for publication). Briefly, this E1$^+$E4$^+$ cell line can be used to produce viral stocks of both an E1$^+$E4$^-$ or an E1$^-$E4$^+$ viral deletant. Importantly, this cell line can sustain cellular confluency long enough so that viral plaques are evidenced following infection with either deletant. Interestingly, E4 expression from this minimal expression unit is also inducible, because plaque formation is apparent only under induction conditions. The efficient dual transcomplementation of both E1 and E4 functions by this cell line is further demonstrated by the construction and plaque purification of several different lacZ(E1$^-$)E4$^-$ third-generation adenovectors (Fig. 5). Extensive molecular and phenotypic characterization of these β-galactosidase-expressing, doubly-deleted adenovectors also demonstrated that they were not contaminated with any E4$^+$ helper virus (Fig. 6, panel B). This result was indeed predictible, since the emergence of an E4$^+$ virus by homologous recombination between the cellular genome and the doubly-deleted adenovectors is basically impossible according to either of the theoretical schemes described in panel A or B of Fig. 2: 1) there are no overlapping sequences between the chromosomal and viral sequences located upstream of ORF6 (see Fig. 3), so a double homologous recombination event on both sides of the integrated E4 functional unit is impossible, and 2) the right adenoviral ITR has not been included in the integrated MMTV/E4 (ORF6 + ORF7) unit, so a single recombination event between the chromosomal and viral sequences

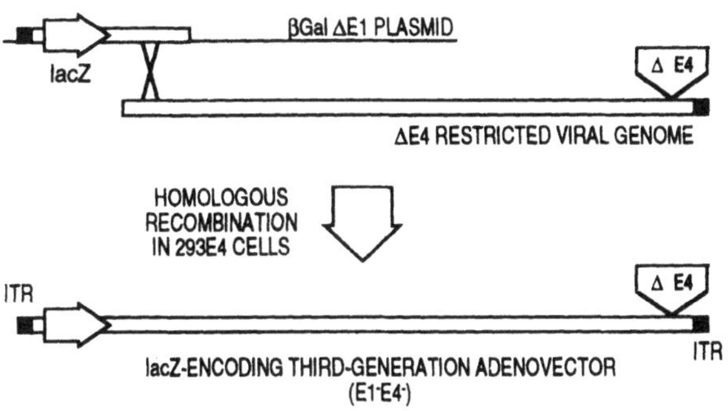

Fig. 5. Strategy for the construction of a lacZ (E1$^-$)E4$^-$ third-generation adenovector in the 293E4 cell line. Note that any β-galactosidase-expressing virus must be generated by homologous recombination between the lacZ-encoding plasmid and the E4$^-$ viral genome

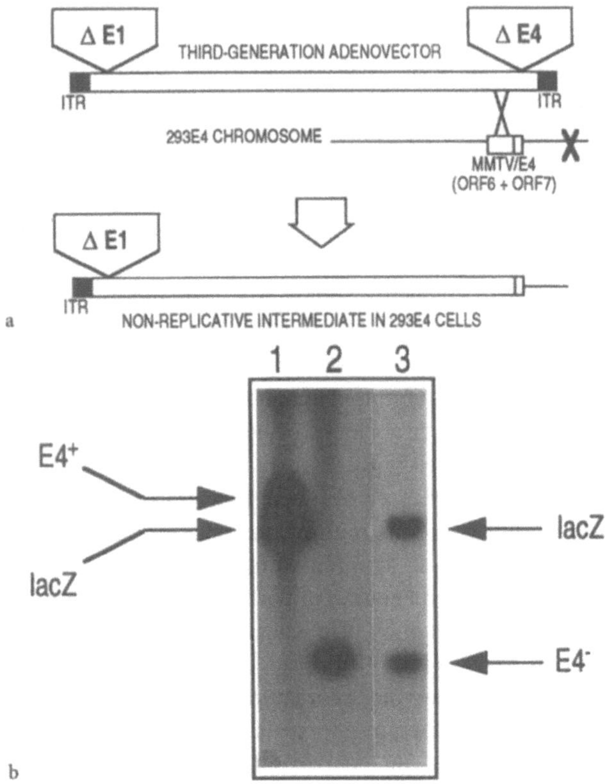

Fig. 6. **a** Homologous recombination between the E4 (*ORF6*+ *ORF7*) sequences from the 293E4 cell line and the backbone of E1$^-$E4$^-$ adenovectors cannot regenerate a replicative E4$^+$ adenovirus **b** absence of E4$^+$ helper virus as detected by Southern analysis of NdeI-restricted viral genomes; *lane* 1 viral DNA from virus RSVβGal (Stratford-Perricaudet et al. 1992); *lane* 2 viral DNA from the E1$^+$E4$^-$ deletant Ad5dl1011 (Bridge et al. 1993); *lane* 3 viral DNA from the E1$^-$E4$^-$ adenovector harboring the dl1011 deletion. A radiolabeled probe encompassing the Ad5 ITR was used

located downstream of ORF7, followed by a putative breakage of the 293E4 chromosome, would generate a lacZ(E1$^-$)E4$^+$ adenoviral intermediate that is unable to replicate, as both ITRs are required for viral propagation (Fig. 6, panel A). This also implies that the further emergence of an E1$^+$E4$^+$ RCA is impossible.

Canine Adenovectors

As demonstrated by the ability of dl808 to rescue (propagate) the E4$^+$ minivirus derived from plasmid pE4Gal, transcomplementation of any defective human adenovectors might occur following in vivo co-expression with wild-type human adenoviruses. It is relevant to recall that integrated adenoviral sequences have been detected in nasal and bronchial samples from a small percentage of untreated individuals (Eissa et al. 1994). Although no RCA could be recovered from these positive samples, this observation raises the possibility of adenovector transcomplementation and the possible emergence of RCA after in vivo recombination following gene transfer in certain individuals, especially when the target cells are part of the adenovirus ecological niche. To avoid narrowing the use of adenovectors in human gene therapy, we are exploring the potential of heterospecific (non-human) adenoviral vectors for human gene therapy. Adenoviral infections occur in a variety of animals, and most adenoviruses are only pathogenic within their specific host. For example, most cells are not permissive for heterospecific adenoviruses. Although they can be infected, they do not allow viral growth, apparently because of poor viral late protein synthesis as exemplified for human Ad2 infection in CV-1 (but not VERO) monkey cells (Anderson and Klessig 1983). The use of a type-2 canine (CAV2) adenovector actually adds an extra safety factor for AMGT in man because human and canine adenoviruses are quite distant on the evolutionary scale (Bailey and Mautner 1994). In fact, CAV2 does not replicate in cultured primary human cells and is apparently not rescued by coinfection with a wild-type human adenovirus. A first-generation CAV2 adenovector expressing the lacZ gene was therefore constructed to document its behavior following in vivo gene transfer (B. Klonjkowski, unpublished observation).

Conclusions

Adenovectors have a tremendous potential for direct gene transfer into humans because they can deliver and express a given therapeutic gene, with no requirement for cell proliferation, and in a large range of somatic cells. Since the first successful somatic gene therapy of a hepatic enzyme deficiency in an animal model was achieved with an adenovirus (Stratford-Perricaudet et al. 1990), recombinant adenoviral vectors have been used to transfer genes in a number of different cell types in vitro and in vivo (for a review, see Kozarsky and Wilson 1993; Trapnell and Gorziglia 1994). Adenoviruses now emerge as potent vectors with a promising future. Specifically, it is now clear that their successful administration directly in vivo in a wide variety of post-mitotic cells opens up the way to new therapeutic strategies.

Following gene transfer, adenovectors remain extrachromosomal once inside the nucleus (they are often mislabeled as episomal although they do not replicate in most human cells). The copy number of the viral genome is reduced during subsequent cell divisions and the adenovectors are eventually lost. Therefore, long-term correction of a genetic defect within proliferative cells is certainly not the target of choice for today's AMGT, because many administrations would be required and adenoviruses are immunogenic particles. These limitations are further supported by the observation that gene transfer efficacy after repeated administration is inversely correlated with the host immunological response to the vector (Yei et al. 1994a).

To avoid exposing the adenovectors to the host neutralizing antibodies, we are currently adapting for AMGT the "neo-organ technology" described for RMGT (Moullier et al. 1993). Briefly, this technology is based on the ex vivo transduction of defined target cells (usually fibroblasts), which are then embedded in a collagen/synthetic polymer network before subsequent, subcutaneous or intraperitoneal implantation. Importantly, neo-organs become highly vascularized enabling this technology to be used for a range of pathological conditions requiring the systemic delivery of secreted therapeutic gene products. The safety features of AMGT are significantly enhanced when using this approach, because it is possible to 1) assay the level of transgene expression before implantation and hence have a greater degree of control; 2) use considerably less adenovectors compared with what is needed for intravenous injection for the same physiological response (Descamps et al. 1995); 3) avoid expression and transduction in undesirable tissues, or in certain subsets of cells exhibiting E1a-like activities; and 4) remove the neo-organ if needed.

The lack of cell surface MHC-I expression may prove useful in gene therapy indications requiring multiple administrations, since an immune response directed towards the virally infected cells could be avoided or significantly reduced. In this context, the expression of particular E3 genes within the vector backbone should certainly be discussed on a case-to-case basis. For example, the constitutive expression of the E3 gp19K protein within an $E1^-$ vector backbone has been shown to reduce the immunogenicity of the infected cells (Lee et al. 1995).

We believe that clinical benefit will first be obtained following AMGT in immunodepressed individuals, and perhaps in immunopriviledged organs (e.g., retina). In this context, the acute treatment of acquired pathologies such as cancer is particularly attractive, especially when a localized intratumoral injection of the vectors is possible. A "metabolic cooperation" (also called "bystander effect") has also been documented following transfer of the herpes simplex thymidine kinase gene and subsequent gancyclovir administration (Moolten 1986). Thus transduction of the whole target cell population is not an absolute requirement to achieve a clinical benefit.

The remaining transcription units of $E1^-E4^-$ adenovectors, and especially the LTU which encodes most viral structural proteins, are likely to be quite silent following gene transfer. This is expected because E1 and E4 are both required for maximal in vitro induction of the E2A promoter (Babiss 1989; Boeuf et al. 1990; Reichel et al. 1989), and because the DBP that is normally expressed from this promoter is itself involved in a 15-fold transactivation of the LTU, at least in vitro (Chang and Shenk 1990; Zijderveld et al. 1994). However, the availability of highly deleted adenovectors will help, but not solve, the problem of their possible in vivo transcomplementation by naturally occurring (human) adenoviruses. Therefore, the development of a heterospecific adenovector (e.g., canine) will undoubtedly prove useful in the design of safer adenovectors for human gene therapy.

Since Edward Tatum proposed almost 30 years ago that it could "be anticipated that viruses will be effectively used for man's benefit, in theoretical studies in somatic-cell genetics and possibly in genetic therapy..." (Tatum 1966), significant progress has been made in the design of safer vectors for direct gene transfer. However, although we believe that DNA is the ultimate drug, the assessment of its full therapeutic potential is nowhere near the end point. Viruses are Nature's own solution to the gene transfer problem, and we are quite confident that their recombinant derivatives will eventually be widely used.

Acknowledgements. We thank Dr. Gary Ketner (Baltimore, MD, USA) for providing us with many adenoviral E4 deletants and the W162 cell line. We also thank Dr. P. Boulanger (Montpellier, France) for providing us with antibodies raised against the adenoviral fiber protein, Dr. T. Shenk (Princeton, NJ, USA) and Dr. P. Hearing (Stony Brook, NY, USA) for their kind gift of antibodies raised against the E4 gene products. We are also grateful to A. Brie, A. Gillardeaux, M. Lepeut and I. Mahfouz for technical assistance, and all members of the laboratory for helpful discussions. E.V. was supported with a grant from the French BioAvenir multidisciplinary program.

References

Alexander IE, Russel DW, Miller AD (1994) DNA-damaging agents greatly increase the transduction of nondividing cells by adeno-associated virus vectors. J Virol 68: 8282–8287

Anderson KP, Klessig DF (1983) Posttranscriptional block to synthesis of a human adenovirus capsid protein in abortively infected monkey cells. J Mol Applied Genet 2: 31–43

Anderson WF (1994) Making clinical grade gene therapy vectors. Human Gene Ther 5: 925–926

Babiss LE (1989) The cellular transcription factor E2f requires viral E1A and E4 gene products for increased DNA-binding activity and functions to stimulate adenovirus E2A gene expression. J Virol 63: 2709–2717

Bai M, Campisi L, Freimuth P (1994) Vitronectin receptor antibodies inhibit infection of HeLùa and A549 cells by adenovirus type 12 but not by adenovirus type 2. J Virol 68: 5925–5932

Bailey A, Mautner V (1994) Phylogenetic relationships among adenovirus serotypes. Virology 205: 438–452

Beier DC, Cox JH, Vining DR, Cresswell P, Engelhard VH (1994) Association of human class I MHC alleles with the adenovirus E3/19K protein. J Immunol 152: 3862–3872

Bett AJ, Prevec L, Graham FL (1993) Packaging capacity and stability of human adenovirus type 5 vectors. J Virol 67: 5911–5921

Boeuf H, Reimund B, Jansen-Durr P, Kédinger C (1990) Differential activation of the E2F transcription factor by the adenovirus E1a and EIV products in F9 cells. Proc Natl Acad Sci USA 87: 1782–1786

Bridge E, Ketner G (1989) Redundant control of adenovirus late gene expression by early region 4. J Virol 63: 631–638

Bridge E, Ketner G (1990) Interaction of adenoviral E4 and E1b products in late gene expression. Virology 174: 345–353

Bridge E, Medghalchi S, Ubol S, Leesong M, Ketner G (1993) Adenovirus early region 4 and viral DNA synthesis. Virology 193: 794–801

Challberg SS, Ketner G (1981) Deletion mutants of adenovirus 2: isolation and initial characterization of virus carrying mutation near the right end of the viral genome. Virology 114: 196–209

Chang L-S, Shenk T (1990) The adenovirus DNA-binding protein stimulates the rate of transcription directed by adenovirus and adeno-associated virus promoters. J Virol 64: 2103–2109
Crystal RG, McElvaney NG, Rosenfeld MA, Chu C, Mastrangeli A, Hay JG, Brody SL, Jaffe HA, Eissa NT, Danel C (1994) Administration of an adenovirus containing the human CFTR cDNA to the respiratory tract of individuals with cystic fibrosis. Nature Genet 8: 42–51
Culver KW, Anderson WF, Blaese RM (1991) Lymphocyte gene therapy. Human Gene Ther 2: 107–109
Descamps V, Blumenfeld N, Perricaudet M, Beuzard Y, Kremer EJ (1995) Adenoviral-organoids directing systemic expression of erythropoietin in mice. Gene Ther, in press
Dix I, Leppard KN (1993) Regulated splicing of adenovirus type 5 E4 transcripts and regulated cytoplasmic accumulation of E4 mRNA. J Virol 67: 3226–3231
Eissa NT, Chu C-S, Danel C, Crystal RG (1994) Evaluation of the respiratory epithelium of normal and individuals with cystic fibrosis for the presence of adenovirus E1a sequences relevant to the use of E1a-adenovirus vectors for gene therapy for the respiratory manifestations of cystic fibrosis. Human Gene Ther 5: 1105–1114
Engelhardt JF, Yang Y, Stratford-Perricaudet LD, Allen ED, Kozarsky K, Perricaudet M, Yankaskas JR, Wilson JM (1993) Direct gene transfer of human CFTR into human bronchial epithelia of xenografts with E1-deleted adenoviruses. Nature Genet 4: 27–34
Engelhardt JF, Ye X, Doranz B, Wilson JM (1994a) Ablation of E2A in recombinant adenoviruses improves transgene persistence and decreases inflammatory response in mouse liver. Proc Natl Acad Sci USA 91: 6196–6200
Engelhardt JF, Litzky L, Wilson JM (1994b) Prolonged transgene expression in cotton rat lung with recombinant adenoviruses defective in E2a. Human Gene Ther 5: 1217–1229
Falgout B, Ketner G (1987) Adenovirus early region 4 is required for efficient virus particle assembly. J Virol 61: 3759–3768
Ferry N, Duplessis O, Houssin D, Danos O, Heard J-M (1991) Retroviral-mediated gene transfer into hepatocytes in vivo. Proc Natl Acad Sci USA 88: 8377–8381
Flomenberg P, Szmulewicz J, Gutierrez E, Lupatkin H (1992) Role of the adenovirus E3-19K conserved region in binding major histocompatibility complex class I molecules. J Virol 66: 4778–4783
Flotte TR, Afione SA, Conrad C, McGrath SA, Solow R, Oka H, Zeitlin PL, Guggino WB, Carter BJ (1993) Stable in vivo expression of the cystic fibrosis transmembrane conductance regulator with an adeno-associated virus vector. Proc Natl Acad Sci USA 90: 10613–10617
Freeman SM, Whartenby KA, Abraham GN, Zwiebel JA (1993) Clinical trials in gene therapy. Adv. Drug Delivery Rev 12: 169–183
Giraud C, Winocour E, Berns KI (1994) Site-specific integration by adeno-associated virus is directed by a cellular DNA sequence. Proc Natl Acad Sci USA 91: 10039–10043
Gooding LR, Ranheim TS, Tollefson AE, Brady HA, Wold WSM (1991) The 10,400- and 14,500-dalton proteins encoded by region E3 of adenovirus function together to protect many but not all mouse cell lines against lysis by tumor necrosis factor. J Virol 65: 4114–4123
Graham FL, Smiley J, Russel WC, Nairn R (1977) Characteristics of a human cell line transformed by DNA from human adenovirus type 5. J Gen Virol 36: 59–72
Halbert DN, Cutt JR, Shenk T (1985) Adenovirus early region 4 encodes functions required for efficient DNA replication, late gene expression, and host cell shutoff. J Virol 56: 250–257
Hemstrom C, Nordqvist K, Pettersson U, Virtanen A (1988) Gene product of region E4 of adenovirus type 5 modulates accumulation of certain viral polypeptides. J Virol 62: 3258–3264
Herz J, Gerard RD (1993) Adenovirus-mediated transfer of low density lipoprotein receptor gene acutely accelarates cholesterol clearance in normal mice. Proc Natl Acad Sci USA 90: 2812–2816

Horwitz MS (1990) Adenoviruses. In: Virology, Fields BN, Knipe DM et al. (eds), Raven Press, New York, pp 1723–1740
Huang MM, Hearing P (1989) Adenovirus early region 4 encodes two gene products with redundant effects in lytic infection. J Virol 63: 2605–2615
Imperiale MJ, Kao H-T, Feldman LT, Nevins JR, Strickland S (1984) Common control of the heat shock gene and early adenovirus genes: evidence for a cellular E1A-like activity. Mol Cell Biol 4: 867–874
Kotani H, Newton III PB, Zhang S, Chiang YL, Otto E, Weaver L, Blaese RM, Anderson WF, McGarrity GJ (1994) Improved methods of retroviral vector transduction and production for gene therapy. Human Gene Ther 5: 19–28
Kotin RM (1994) Prospects for the use of adeno-associated virus as a vector for human gene therapy. Human Gene Ther 5: 793–801
Kozarsky KF, Wilson JM (1993) Gene therapy: adenovirus vectors. Curr Opin Genet Dev 3: 499–503
La Thangue NB (1994) DRTF1/E2F: an expanding family of heterodimeric transcription factors implicated in cell-cycle control. Trends Biochem Sci 19: 108–114
La Thangue NB, Rigby PWJ (1987) An adenovirus E1A-like transcription factor is regulated during the differentiation of murine embryonal carcinoma stem cells. Cell 49: 507–513
Lee MG, Abina MA, Haddada H, Perricaudet M (1995) The constitutive expression of the immuno-modulatory gp19k protein in E1-,E3- adenoviral vectors strongly reduces the host cytotoxic T cell response against the vector. Gene Ther, in Press
Leonard CJ, Berns KI (1994) Adeno-associated virus type 2: a latent life cycle. Progress Nucl Acid Res Mol Biol **48**: 29–52
Lochmuller H, Jani A, Huard J, Prescott S, Simoneau M, Massie B, Karpati G, Acsadi G (1994) Emergence of early region 1-containing replication-competent adenovirus in stocks of replication-defective adenovirus recombinants (DE1 + DE3) during multiple passages in 293 cells. Human Gene Ther 5: 1485–1491
Mathias P, Wickham T, Moore M, Nemerow G (1994) Multiple adenovirus serotypes use α_v integrins for infection. J Virol 68: 6811–6814
Moolten FL (1986) Tumor chemosensitivity conferred by inserted Herpes thymidine kinase genes: Paradigm for a prospective cancer control strategy. Cancer Res 46: 5276–5281
Moullier P, Marechal V, Danos O, Heard J (1993) Continuous systemic secretion of a lysosomal enzyme by genetically modified mouse skin fibroblasts. Transplantation 56: 427–432
Muller U, Kleinberger T, Shenk T (1992) Adenovirus E4Orf4 protein reduces phosphorylation of c-Fos and E1A proteins while simultaneously reducing the level of AP1. J Virol 66: 5867–5878
Nemerow GR, Cheresh DA, Wickham TJ (1993) Adenovirus entry into host cells: a role for α_V integrins. Trends Cell Biol 4: 52–55
Nordqvist K, Akusjarvi G (1990) Adenovirus early region 4 stimulates mRNA accumulation via 5′ introns. Proc Natl Acad Sci USA 87: 9543–9547
Nordqvist K, Ohman K, Akusjarvi G (1994) Human adenovirus encodes two proteins which have opposite effects on accumulation of alternatively spliced mRNAs. Mol Cell Biol 14: 437–445
Ohman K, Nordqvist K, Akusjarvi G (1993) Two adenovirus proteins with redundant activities in virus growth facilitates tripartite leader mRNA accumulation. Virology 194: 50–58
Onclercq R, Gilardi P, Lavenu A, Cremisi C (1988) c-myc products trans-activate the adenovirus E4 promoter in EC stem cells by using the same target sequence as E1A products. J Virol 62: 4533–4537
Otto E, Jones-Trower A, Vanin EF, Stambaugh K, Mueller SN, Anderson WF, McGarrity GJ (1994) Characterization of a replication-competent retrovirus resulting from recombination of Packaging and vector sequences. Human Gene Ther 5: 567–575

Payne GS, Bishop JM, Varmus HE (1982) Multiple arrangements of viral DNA and an activated host oncogene in bursal lymphomas. Nature 295: 209–214
Pilder S, Moore M, Logan J, Shenk T (1986) The adenovirus E1B-55K transforming polypeptide modulates transport or cytoplasmic stabilization of viral and host cell mRNAs. Mol Cell Biol 6: 470–476
Ragot T, Vincent N, Chafey P, Vigne E, Gilgenkrantz H, Couton D, Cartaud J, Briand P, Kaplan J-C, Perricaudet M, Kahn A (1993) Efficient adenovirus-mediated transfer of a human minidystrophin gene to skeletal muscle of mdx mice. Nature 361: 647–650
Reichel R, Neill SD, Kovesdi I, Simon MC, Raychaudhuri P, Nevins JR (1989) The adenovirus E4 gene, in addition to the E1A gene, is important for trans-activation of E2 transcription and for E2F activation. J Virol 63: 3643–3650
Russell DW, Miller AD, Alexander IE (1994) Adeno-associated virus vectors preferentially transduce cells in S phase. Proc Natl Acad Sci USA 91: 8915–8919
Samulski RJ, Zhu X, Xiao X, Brook JD, Housman DE, Epstein N, Hunter LA (1991) Targeted integration of adeno-associated virus (AAV) into human chromosome 19. EMBO J 10: 3941–3950
Simon RH, Engelhardt JF, Yang Y, Zepeda M, Weber-Pendleton S, Grossman M, Wilson JM (1993) Adenovirus-mediated transfer of the CFTR gene to lung of nonhuman primates: toxicity study. Human Gene Ther 4: 771–780
Spergel JM, Chen-Kiang S (1991) Il-6 enhances a cellular activity which functionally substitutes for E1A in trans-activation. Proc Natl Acad Sci USA 88: 6472–6476
Spergel JM, Hsu W, Akira S, Thimmappaya B, Kishimoto T, Chen-Kiang S (1992) NF-IL6, a member of the C/EBP family, regulates E1A- responsive promoters in the absence of E1A. J Virol 66: 1021–1030
Stratford-Perricaudet LD, Levrero M, Chasse J-F, Perricaudet M, Briand P (1990) Evaluation of the transfer and expression in mice of an enzyme-encoding gene using a human adenovirus vector. Human Gene Ther 1: 241–256
Stratford-Perricaudet LD, Makeh I, Perricaudet M, Briand P (1992) Widespread long-term gene transfer to mouse skeletal muscles and Heart. J Clin Invest 90: 626–630
Tatum EL (1966) Molecular biology, nucleic adids and the future of medecine. Perspect Biol Med 10: 19–32
Tian M, Maniatis T (1993) A splicing enhancer complex controls alternative splicing of doublesex pre-mRNA. Cell 74: 105–114
Trapnell BC, Gorziglia M (1994) Gene therapy using adenoviral vectors. Curr Opin Biotech 5: 617–625
Valcarcel J, Singh R, Zamore PD, Green MR (1993) The protein Sex-lethal antagonizes the splicing factor U2AF to regulate alternative splicing of transformer pre-mRNA. Nature 362: 171–175
Van der Vliet PC, Sussenbach JS (1975) An adenovirus type 5 gene function required for initiation of viral DNA replication. Virology 67: 415–426
Virtanen A, Gilardi P, Naslund A, LeMoullec JM, Pettersson U, Perricaudet M (1984) mRNAs from human adenovirus 2 early region 4. J Virol 51: 822–831
Walsh CE, Nienhuis AW, Samulski RJ, Brown MG, Miller JL, Young NS, Liu JM (1994) Phenotypic correction of Fanconi anemia in human hematopoietic cells with a recombinant adeno-associated virus vector. J Clin Invest 94: 1440–1448
Weinberg DH, Ketner G (1983) A cell line that supports the growth of a defective early region 4 deletion mutant of human adenovirus type 2. Proc Natl Acad Sci USA 80: 5383–5386
Weinberg DH, Ketner G (1986) Adenoviral early region 4 is required for efficient viral DNA replication and for late gene expression. J Virol 57: 833–838
Weinberg RA (1989) Oncogenes, antioncogenes, and the molecular bases of multistep carcinogenesis. Cancer Res 49: 3713–3721

Weitzman MD, Kyostio SRM, Kotin RM, Owens RA (1994) Adeno-associated virus (AAV) Rep proteins mediate complex formation between AAV DNA and its integration site in human DNA. Proc Natl Acad Sci USA 91: 5808–5812

Wold WSM, Gooding LR (1991) Region E3 of adenovirus: a cassette of genes involved in host immunosurveillance and virus-cell interactions. Virology 184: 1–8

Yang Y, Nunes FA, Berencsi K, Furth EE, Gonczol E, Wilson JM (1994a) Cellular immunity to viral antigens limits E1-deleted adenoviruses for gene therapy. Proc Natl Adac Sci USA 91: 4407–4411

Yang Y, Nunes FA, Berencsi K, Gonczol E, Engelhardt JF, Wilson JM (1994b) Inactivation of E2a in recombinant adenoviruses improves the prospect for gene therapy in cystic fibrosis. Nature Genet 7: 362–369

Yei S, Mittereder N, Tank K, O'Sullivan C, Trapnell BC (1994a) Adenovirus-mediated gene transfer for cystic fibrosis: quantitative evaluation of repeated in vivo vector administration to the lung. Gene Ther 1: 192–200

Yei S, Mittereder N, Wert S, Whitsett JA, Wilmott RW, Trapnell BC (1994b) In vivo evaluation of the safety of adenovirus mediated transfer of the human cystic fibrosis transmembrane conductance regulator cDNA to the lung. Human Gene Ther 5: 731–744

Zapner J, Petersen DM, Puga AP, Graham SM, Couture LA, Keyes LD, Lukason MJ, St George JA, Gregory RJ, Smith AE, Welsh MJ (1994) Safety and efficacy of repetitive adenovirus-mediated transfer of CFTR cDNA to airway epithelia of primates and cotton rats. Nature Genet 6: 75–83

Zijderveld DC, d'Adda di Fagagna F, Giacca M, Marc Timmers HT, Van der Vliet P (1994) Stimulation of the adenovirus major late promoter in vitro by transcription factor USF is enhanced by the adenovirus DNA binding protein. J Virol 68: 8288–8295

Towards an Ethics of 'Complexity' for 'Common' Diseases?

B.M. KNOPPERS

Human genetics is not just personal but familial, social and universal. It affects whole populations as well as future generations. It has its own myths, its own meanings, its own imagery. No longer limited to finding specific genes for single gene disorders, it is increasingly becoming the human genetics of common diseases, the latter described in the terminology of risk, susceptibility, probability, and predisposition. Besides individuals and families, these genetic factors found in common diseases concern communities, that is, populations or subpopulations of persons around the world regrouped "genetically" by disease, ethnic origin, race, gender, age, or region. Indeed, genetic families once limited to genealogies and pedigree studies for single gene diseases will gradually be replaced by new "families" whose membership resides in their at-risk status for common diseases such as breast cancer or heart disease. Moreover, although many factors contribute to these common diseases, genetic factors could be the unifying link in these new extended families. Do we simply multiply the ethical, legal and social issues already known to affect individuals and then factor in this new information? Or does genetics in collectivities, in these communities, in these populations, raise different questions and concerns requiring a different ethic - a complex ethic or an ethics of complexity for common diseases (for references, see Roy et al. 1995)? To answer that question, we must first examine the current social representations of known, monogenic disorders as found in the ethical-legal literature (Knoppers et al. 1991-1994) before turning to complexity theory and, finally, to the proposal of new ethical principles (Knoppers 1991).

The language of this ethical-legal literature is both polemic and phobic. Perhaps this is due to the absence of an open and factually based public debate. The myriad of social representations of human genetics found in the medical, legal and bioethics literature is haunted both by the fatalism of lethal, single gene diseases and by the myths of ancient monsters and modern machines - to the extent that the presumed social and legal "malformations" now seem to equal the genetic ones! Indeed, the literature contains 10 sets of social representations surrounding human genetics. If these representations ("imaginaires") are given further credibility or emphasis, they will serve to

K. Berg, V. Boulyjenkov, Y. Christen (Eds.)
Genetic Approaches to Noncommunicable Diseases

exacerbate these current social and legal "malformations" when applied to common diseases of which genetics is only a factor. The following 10 polarized sets of social versus scientific interpretations of the advances in human genetics are presented here in a heuristic typology, that of either social predestinations or of scientific liberation (Knopper 1994).

The first representation found in the literature is that associated with the popular understanding of the term mutation. The term is often confused with the pejorative sense of "mutant" and has a potentially dangerous social impact. Failure to understand that we all carry four to seven deleterious genes, and that even single gene disorders are complex and highly variable in their timing and intensity, underscores this perception. As concerns genetic factors in common diseases, the public would be better served by terms like probability, susceptibility and predisposition and by the notions of intervention, adaptation and adjustment, to say nothing of the scientific language of variation.

The second representation is that of human manipulation and its associated monster mythology. This term is inaccurate in that most successful genetic "manipulation" (e.g., somatic therapy for ADA deficiency) is actually closer to transplantation than it is to the horrendous images of the "Frankenstein factor" or, more recently, of "Jurassic Park".

The third representation is that of elimination. Natural elimination is part of the human condition, as in the case of spontaneous abortion. There is no doubt that the procedure of biopsy and diagnosis of an embryo, prior to the implanation of only those embryos found to be free of the disorder in question, is a form of elimination. Deliberate elimination, however, is not an explicit research goal of the Human Genome Project and this technique is not applied outside of those single gene disorders with a high degree of morbidity and mortality. Rather, the actual situation is closer to the fourth representation, that of selection. Indeed, in contrast to the previous representations, this term is not used wholly inaccurately, because preimplanation diagnosis is a form of selection. Yet, like prenatal diagnosis, it also involves the provision of information and the possibility of reproductive choice for at-risk couples who may not have conceived otherwise- a form of validation and prevention.

The fifth representation is that of genetic identification. Although this has nothing to do with medical genetics, it uses the technologies provided by molecular biology. Again, this term must be qualified, since DNA typing permits greater accuracy in profiling serious offenders, missing children and putative fathers. It provides the ultimate individualization of the person in its proof of personal uniqueness. However, at the same time that it provides greater certainty, it also calls into question fundamental human rights, such as privacy and the interests of the child. Scientific oversight of the quality of testing laboratories, and judicial oversight of how samples are obtained and for what purpose(s),are issues that have yet to be discussed in many countries.

The sixth representation is that of information. On the one hand, it could be argued that too much information leads to the "geneticization" of a society whose members are unwillingly or unknowingly subjected to technological imperative (Lippman 1991). On the other hand, the increased availability of genetic information can prevent human suffering, provided that individual decisions remain personal and free, as opposed to political or economic. The introduction of technologies without medical education, public information and discussion could tip this balance toward the negative, however!

The seventh representation is that of discrimination. Could it not be said, however, that any discrimination in insurance or employment due to the availability of genetic information about individuals stems more from the erroneous understanding of genetic information and of genetic disease than from genetic knowledge itself? One recent Quebec case (Audet v. L' Industrielle-Alliance; Billings in Gostin 1991; Billings et al. 1992) illustrates that discrimination in insurance can stem not only from the usual social or systemic sources, but also from the failure of the court to obtain accurate expert testimony about the extreme variability of the genetic condition in question (myotonic dystrophy). In other words, the court failed to understand that one can have a genetic disease without expressing it. When the insurance applicant wrote "no" on a form to the question concerning the presence of any "anomalies" the answer was not false. Cognizant of the forms of the disease in his family and in the region, he was well aware that you can "have" it without "having" it. His widow, however, was deprived of all insurance benefits.

The eighth representation is that of stigmatization - "diseased" by association. Here, the uneasy formula surfaces that equates the person (or population or group) with the "defective" gene. The gene becomes the disease, the disease is the person and thus, the person becomes the gene. In the same way, certain ethnic groups, populations, regions or countries that have profited from extensive genealogical and genetic research or where the incidence of certain monogenic diseases is high now may find themselves labelled in the press as genetically defective.

The ninth representation is that of commercialization. The trend of making the human body a commodity, which has gone as far as assigning an actual or potential economic value to parts and even cells, constitutes an affront to human dignity and a danger to research. Take, for example, the famous Moore case (Moore v. Regents of University of California 1988) in California, where a Coca-Cola salesman with a rare form of leukemia discovered that researchers had developed and patented a pharmacological product using his cells without his knowledge or consent. Although the higher court did not definitively pronounce itself on the property issue, it did say that the researchers should have disclosed their conflict of interest and obtained an informed consent.

It is clear from this brief journalistic typology of possible and actual representations that there is a need for increased understanding and information about the new human genetics and the Human Genome Project. In the current North American context, with its emphasis on rights and conflicts of rights, the absence of knowledge could well lead to the emergence of a tenth representation - that of litigation. Claims of genetic malpractice or or wrongful birth or life could make the new human genetics the source of legal "eugenics!" To even talk of the "genetic right" to be conceived, to be born with or without certain genetic conditions, to be manipulated or free from manipulation, is dangerous territory indeed. It is here that state efforts should include a legislated prohibition against quality of life suits against parents.

These myths, these perceptions and representations, illustrate the possible ethical and legal "malformations" and misconceptions that come when we neglect to correctly inform and involve the patients, families, communities and populations in the debate over the directions of human genetics. Now that new at-risk families and communities will be discovered and perhaps come forward for testing for genetic factors in common diseases, these problems will not only increase in magnitude but also in complexity; hence, the priority of education.

As early as 1991, an article on the social geography of human genome mapping (Knoppers and Laberge 1991) argued against a polarized, isometric and linear view of human genetics and in favor of a stereoscopic, integrated approach. To do otherwise underestimates the variability and complexity of humans and their genes. Likewise, a recent article (Strohman 1994) argues against the extension of monogenic logic based on a reductionist approach to understanding disease, in favor of an interactive, dynamic understanding of the two informational systems found in cells, the genetic and epigenetic. According to the author the system is thus "a determinative chaotic system open to new approaches that combine linear genetic with non-linear complex system (epigenetic) analysis".... At cellular and higher levels... evolution has worked to select not just single genes but integrated behavior or generic patterns of response at all levels of biological organization. These patterns cannot be seen by linear analysis. It is at this level that theory of complex systems might prove useful" (Strohman 1994).

There is currently no single, comprehensive definition of complexity that is generally acceptable to all persons studying complex systems in the natural and social sciences. Nevertheless, many systems, be they physico-chemical, biological, or social - for example, economics, communications, ethics-share and display in common a core of properties that are recognized across disciplines as representing complexity.

These complex systems have been described as:

"sharing five common behaviors that: are proper to the system as a whole and emerge from the multiple and diverse mutual interactions of the system's many component parts, agents, or events (the property of emergence); emerge across successively higher organizations of interacting components within the system - lower through intermediate to higher the behavior of the system as a whole (property of hierarchy); cannot be explained in terms of any one component or sub-system within the system (property of non-reducibility); cannot be explained in terms of any single chain of interacting components within the system (property of non-linearty) and are unsubmissive to precise prediction: complex systems do not deliver or guarantee close knowledge of that system's final state (property of being non-deterministic)" (Roy et al. 1995).

Furthermore, according to this approach, the ethical, legal and social issues linked to the use of genomics for multifactorial conditions must be situated on:

"a continuum from lower to higher complexity depending on: the numbers of people afflicted; the multiple institutions and sectors of society that are involved; the multiplicity of preventive and therapeutic approaches that can or should be attempted; the diversity of sources of responsibility linked to the different factors (individual, industrial, societal, etc.), which can lead to a continual shifting of responsibility from one factor to another; the uncertainty of prediction of time of onset, progression, severity of multifactorial diseases due to the dependence of outcome on the interaction of many probabilities; and the levels of difficulty involved in understanding the causation and ways of managing disease" (Roy et al. 1995).

A mechanistic and reductionist view of human genetics cannot do justice to the complex epigenesis of multifactorial conditions. Neither is it sufficient for even the rare, monogenic diseases that are not proving so simple and are defying the traditional linear approaches themselves as new mutations are discovered. What is need then is an ethics of complexity that is adaptive and evolutionary.

A study covering the more than 30 reports, bills and laws on human genetics between 1989 and 1994 (Knoppers and Chandwick 1994) reveals the gradual emergence of five areas of consensus on common principles arising from the need to respect human dignity: privacy, autonomy, equity, justice and qualtiy assurance. This consensus is not that stemming from retrospective generalization, as was the case with reproductive technologies or organ transplantation. In those cases, two decades of incremental case law, professional self-regulation, legislation and sporadic scandals culminated in both the 1991 WHO international transplantation guidelines (World Health Organization 1991) and the future European Bioethics Convention (Council of Europe 1994) Neither is this consensus the result of a deliberate effort towards either uniformization or transnational harmoniz- ation. Rather, it is a natural convergence towards a prospective approach stemming from the realization of the futuristic and universal character of the Human Genome Project (Knoppers and Le Bris 1994).

As a first step in the translation and application of those five common principles in the actual practice of human genetics and, more particularly, in the elaboration of an ethics of complexity for common diseases, I would propose four new ethical principles that may serve to reflect the relationships, risks and responsibilities inherent in human genetics (Knoppers 1991).

At the first level, that of the physician-patient relationship, the ethic of *reciprocity* means the recognition of a partnership as well as openness with respect to goals, research options and alternatives. The giving of information and of DNA places obligations on the physician-researcher to safeguard against misuse and to communicate results and put in place options for follow-up or recall procedures where so desired. At the level of families who have in common their genetic, at-risk status, the ethic of *mutuality* means that it is family members or at-risk communities who solicit and contact family members and communicate genetic risk information among themselves. At the level of the citizen, in relationship to the State, the ethic of *solidarity* means that, in exchange for the participation of individuals, families and populations, special protection against adverse socio-economic consequences should be put in place by the State. Finally, seeing that common genetic diseases cross borders and oceans, creating new genetic families and populations, we may also need to consider that of *universality*, that is our obligations as world citizens to ensure against genetic dumping and exploitation and to promote free exchange and scientific collaboration. Perhaps the adoption and integration of these new principles may mitigate, if not eradicate, the current social and legal malformations that do not afford the necessary respect for the complexity and variability of the human person. Then we can denounce those who dwell in what George Orwell has called, "the huge dump of worn-out metaphors (representations?; author's question) which have lost all evocative power and are merely used because they save people the trouble of inventing phrases for themselves" (Bonnicksen 1994).

References

Audet v. L'Industrielle-Alliance (1990) R.R.A. 500; Gostin L (1991) Genetic discrimination; the use of genetically based diagnostic and prognostic tests by employers and insurers. Am JL Med 17: 109–144

Billings PR, Kohn MA, De Cuevas M, Beckwith J, Alper JS, Natowicz MR (1992) Discrimination as a consequence of genetic testing. Am J Human Genet 50: 476–482

Bonnicksen A (1994) Demystifying germ-line genetics. Politics Life Sci 13: 246–248

Council of Europe (1994) Draft convention for the protection of human rights and dignity of the human being with regard to the application of biology and medicine: bioethics convention

Knoppers BM (1991) Human dignity and genetic heritage. Law Reform Commission of Canada, Ottawa, pp 69–72

Knoppers BM (1994) Issues in genetic research. Communiqué 5: 1–2

Knoppers BM, Chadwick R (1994) The human genome project: under an international ethical microscope. Science 265: 2035–2036

Knoppers BM, Laberge CM (1991) The social geography of human genome mapping. In: Bankowski Z, Capron AM (eds) Genetics, ethics and human values. Proceedings of the XXIVth CIOMS Conference, Geneva, 56

Knoppers BM, Le Bris S (1994) Genetic choices: a paradigm for prospective international ethics? Politics Life Sci 13: 228–229

Knoppers BM, Laberge C, Goulet J, Bourgeault G, Mackay P, Rocher G (1991–1994) Social, legal and ethical issues of genetic epidemiology. Social Sciences and Humanities Research Council of Canada

Lippman A (1991) Prenatal genetic testing and screening: constructing needs and reinforcing inequities: Am JL Med 1: 15–50

Moore v. Regents of the University of California (1988) 249 Cal. Rptr. 494 (Cal. App. 2 Dist); July 9, 1990, Supreme Court of California, 271 Cal Rptr 146

Roy D, Davignon J, Rocher G (1995) Genomics and multifactorial diseases: towards an ethics for complexity (collaborators: Sing C, De Langavant G). CGAT Project 1995. Bioethics Centre, IRC, Montreal

Strohman R (1994) Epigenesis; the missing beat in biotechnology? Bio/Technology 12: 156–164

World Health Organization (1991) Human organ transplantation - a report on developments under the auspices of WHO (1987–1991). Geneva, WHO

Disorders with Complex Inheritance in India: Frequency and Genetic/Environmental Interactions

I.C. VERMA

The Epidemiologic Transition in Developing Countries

Due to the remarkable successes in recent years in the implementation of the immunization program and in the provision of primary health care, India is passing through an epidemiologic transition. It is burdened with not only numerous infectious and nutritional problems, but also with a large number of non-communicable disorders (Ministry of Health and Family Welfare 1994). The non-communicable disorders comprise genetic diseases, cardiovascular disorders, diabetes mellitus, mental illnesses, mental retardation and cancer, all of which have a significant genetic component. Figure 1 depicts the estimated and projected mortality rates for major causes of death in India for 1985, 2000, and 2015. The expected reduction in the number of cases of infectious diseases and the doubling in the number of subjects with cardiovascular disorders and cancer are immediately apparent.

Disorders with Complex Inheritance

Among genetic disorders, the most common in frequency are those that have a "complex" etiology. These disorders are due to the combined effect of many genes interacting with the environment. In recent years, advances in genetic methodologies have not only confirmed the concept of "complex" or "polygenic" inheritance, but have also identified some genetic and environmental components in congenital malformations, coronary artery disease, diabetes mellitus and other disorders (Thomson 1994; Copeman et al. 1994; Romeo and McKusick 1994).

Figure 2 depicts the chronology of presentation of genetic disorders in a life span. Genetic disorders occur in two waves: the first at birth, and the second in adult life. At birth the disorders with complex inheritance manifest as congenital malformations due to defects in morphogenesis. Disorders presenting in adult life are the common disorders affecting the cardiovascular system (coronary artery disease and hypertension), nervous system (psychiatric disorders and epilepsy),

K. Berg, V. Boulyjenkov, Y. Christen (Eds.)
Genetic Approaches to Noncommunicable Diseases

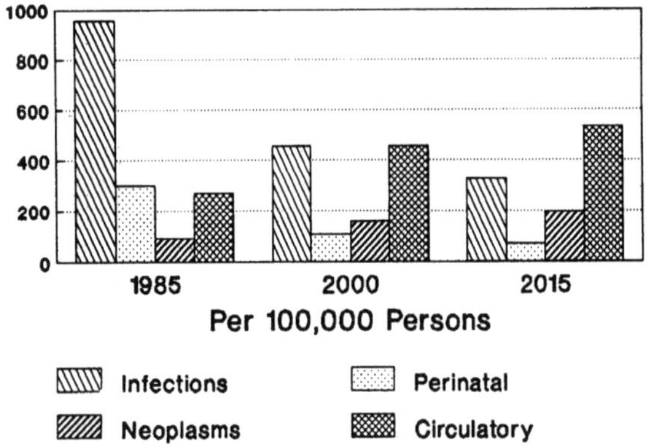

Fig. 1. Estimated (1985) and projected (2000, 2015) mortality rates for major causes of death in India

and endocrine system (diabetes mellitus). The frequency of multifactorial disorders-taken over the whole life span to include malformations at birth and late-onset disorders of middle and old age-is of the order of 60% (Scriver 1992). This appears astonishing, but it is true!

This paper summarizes the prevalence of complex disorders in India, along with their important epidemiologic characteristics.

Congenital Malformations

Neural Tube Defects in India

Congenital malformations were found to have a frequency of 19.4/1000 births in a meta-analysis of all published studies involving 301,987 births in 25 hospitals

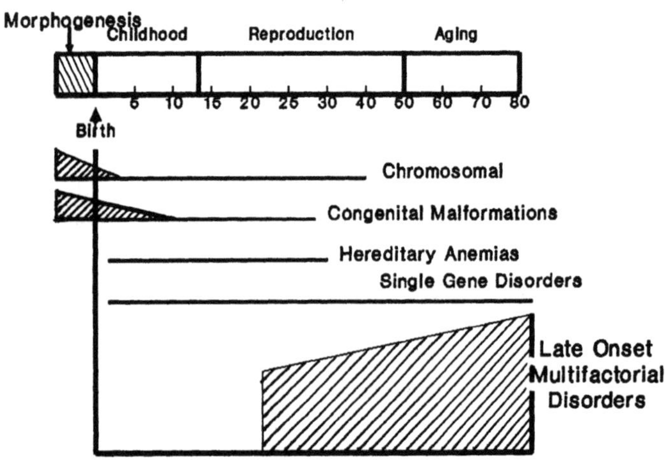

Fig. 2. Chronology of genetic disorders in a life span

(Verma et al. 1990). Among 8,409 stillbirths in the same analysis, the malformation rate was 91/1000. We also carried out a prospective national multi-centric study involving 102,224 births in 18 hospitals using a standardized proforma (Verma and Mehta, unpublished observations). The frequency of total malformations was 16.7/1000. Table 1 sets out the mean frequency of malformations obtained in this study. Neural tube defects (NTD), comprising spinal bifida and anencephaly, were the most common, with a mean frequency of 3.5/1000.

Neural tube defects (spina bifida and anencephaly) are multifactorial in etiology and scientists are beginning to unravel the genetic and environmental factors involved. Indeed, one of the environmental factors has already been identified. In women who had previously had a child affected with a neural tube defect, the administration of folic acid 1 month before pregnancy and continued up to 3 months of pregnancy remarkably reduced (by 70%) the recurrence of NTD in a subsequent pregnancy (Wald et al. 1991). This double-blind, randomized trial confirmed the earlier observations of Smithells et al. (1980), showing the beneficial effect of folic acid supplementation during the periconceptional period. In a more recent study periconceptional administration of folic acid was able to prevent the occurrence of even the first instance of NTD (Czeizel and Dubas 1992). Under the aegis of the Indian Council of Medical Research, a multicentric study was carried out in India to examine the effect of folic acid supplementation in women who had previously given birth to a child with spina bifida/anecephaly. This was a double-blind and placebo-controlled trial. About 470 women were enrolled; 232 were in the vitamin group and 235 in the placebo group. However, only 136 completed the trial in the vitamin group and 143 in the placebo group. Periconceptional folic acid supplementation led to a 45% protective effective against frequency of spontaneous abortions, and a 40.8% protective effect against occurrence of NTD (ICMR Human Genetics Task Force Report 1992).

Genetic and Environmentl Factors in Experimental Spina BiÆda

Malformations are generally considered environmentally induced. However, in a large and carful study by Nelson and Holmes (1989), teratogenic factors could be

Table 1. Mean frequency of congenital malformations in the National Multi-Centric Study [n = 102,224][a]

Disorder	Frequency per 1,000
Neural tube defects (anecephaly ± spina befida)	3.5
Talipes equinovarus	2.3
Polydactyly	1.7
Congential heart disease	1.3
Cleft lip ± cleft palate	1.3
Gastro-intestinal abnormalities	1.2
Hydrocephalus	1.06

[a]Verma and Mehta, unpublished data.

identified only in 3.2% of cases, whereas genetic and multifactorial factors were recognized in 50.7% of cases. Other investigators estimate that genetic predisposition is present in about 90% of human birth defects (Copp 1994).

Data are accumulating that etiologic factors operate through cellular and molecular mechanisms to cause birth defects (Fig. 3). Many genes have been found to regulate fetal development in lower animals. Homologues of most of these genes are being identified in humans (Wilson 1993; Romeo and McKusick 1994). These genes secrete peptide growth factors that act as important signals for developmental events. For example, Gurken gene acts through the transcription factor TGFa in *Drosophila*, TGF β has multiple inductive roles in development, Wnt gene is involved in patterning of nervous system in vertebrates, and Shh gene is involved in dorsoventral patterning of neural tube. PAX genes are developmental control genes encoding transcription factors containing DNA binding paired domain (Strachan and Read 1994). Mutations in three of the nine mouse PAX genes (1, 3, and 9) and two of the nine human PAX genes (3 and 6) are known to cause developmental defects. Mutations in the PAX3 gene have been identified in Waardenberg syndrome type 1, whereas PAX6 mutations have been recognized in several patients with aniridia, as well as in Peters anomaly (a congenital defect of the anterior chamber of the eye). Some genes governing developmental processes in humans are listed in Table 2.

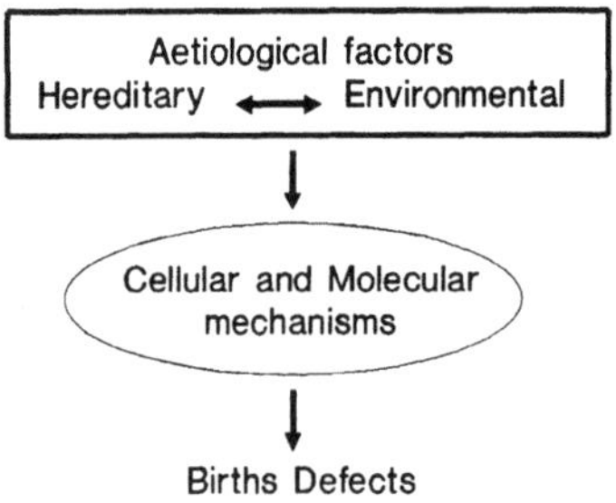

Fig. 3. Pathogenesis of birth defects

Table 2. Genes involved in developmental processes[a]

Gene	Function	Developmental process	Human malformation
Oct-3	Transcription factor	Zygote cleavage	-
Wnt-1	Signal transduction	Axis formation	-
Pax 3	Transcription factor	Segmentation	Waardenberg syndrome
NF-1	Signal transduction	Cell growth	Neurofibromatosis
WT-1	Transcription factor	Differentiation	Genital/kidney defects
Gli 3	Transcription factor	Differentiation	Grieg syndrome
C-Kit	Signal transduction	Cell migration	Piebald trait
Kalig 1	Cell adhesion	Cell migration	Kallman syndrome
Fibrillin	Mechanical integration	Tissue growth	Marfan syndrome

[a]Adapted from Wilson (1993).

Many oncogenes or tumour suppressor genes, which in the mutated form lead to malformations, have been termed as "teratogenes" by Romeo and McKusick (1994). Table 3 summarises the data on the teratogenes known in man.

As NTDs are the commonest malformations observed in developing countries, the experimental approaches used to dissect the environmental and genetic components in their causation will be briefly discussed. The curly tail mouse (ct/ct) is the best-studied animal model of spina bifida (Sellers 1982). Interestingly, the basic defect in the curly tail mouse is not in the neural plate, but in the ventrally located tissues, i.e., the notochord and the hindgut (Copp 1994). This finding has been confirmed by the demonstration that the neural plate closes normally when cultured without the ventral tissues. It has been further demonstrated that the defect originates from a reduced rate of cell proliferation in the notochord and the hindgut. This reduced rate is due to a reduced uptake of transferrin, which leads to a reduced accumulation of newly synthesized hyaluronon in the basement membrane regions beneath the neuroepithelium and a reduced rate of cell proliferation in hindgut endoderm and caudal notochord. This latter process increases the ventral curvature of the caudal region and results in spina bifida. Recently, Newmann et al. (1994) identified a gene on chromosome 4 that mutates to form spina bifida in the curly tail mouse. Possible candidate genes are PAX7, Hspg2 (perlecan), Synd 3 (Syndecan3) Fgr and Cdcl 1. The putative gene would encode a protein involved in the control of cell proliferation in the notochord and hindgut endoderm. At least three modifier genes were also identified, as frequency of NTD varied with three different inbred strains. Two of these genes map to chromosomes 3 and 5. Predisposing genetic constitution (genes) interacts with environmental factors to cause birth defects. For example, on exposure to cortisone C57BL/6, inbred strains of mice develop cleft palate, whereas the Ct strain of mice develop NTD. However, in experimental studies, a number of environmental influences have been identified that lead to spina bifida in the curly tail mouse, such as hyperthermia, valproate, vitamin A, hydroxyurea, mitomycin C, maternal starvation, and inositol deficiency. Methionine is known to protect against development of spina bifida.

Table 3. Teratogenes identified in humans[a]

Gene	Chromosomal location	Congenital anomaly
Pax 3	2q35	Waardenberg syndrome
Pax 6	HP13	Aniridia/Peters anomaly
VHL	3p26-p25	Van Hippel Lindau syndrome
Kit	4q12	Piebaldism
Gli 3	7p 13	Greig cephalopolysyndactyly
RET	10q11.2	Hirschsprung disease
WT 1	11p13	Denys-Drash syndrome
NF1	17q 11.2	Watson syndrome

[a]Adapted from Romeo and McKusick (1994).

One environmental factor identified in the etiology of NTD in humans is folic acid; folic acid supplementation in the periconceptional period prevents the recurrence, or the first occurrence, of NTD. However, the precise mechanism of action of folic acid remains to be elucidated. Is it due to a foliate-related defect in the embryo, or is the action purely pharmacological through administration of folic acid to the mother? Bower et al. (1993) found normal folate uptake or metabolism in mothers of children affected with spina bifida.

Complex Disorders of Late Onset

Table 4 summarises the burden of late-onset disorders in India. These six disorders affect almost 88.8 million persons in India. These "common" disorders are the result of interaction of genetic predisposition with environmental, dietary and life style factors. The evidence for genetic factors is particularly compelling for coronary artery disease, diabetes mellitus and hypertension.

Coronary Artery Disease

It is a common observation that the frequency of coronary artery disease is on the rise in developing countries. It is the highest ranking cause of mortality in middle-aged men in Sri Lanka. Mendis (1992) showed that 50% of Sri Lankan men had a high serum cholesterol level. In India, prevalence of coronary artery disease, based on electrocardiography, was reported to be 65.4/1000 males over the age of 30 in Chandigarh. In a village in Haryana the prevalence was lower, at 22.8/1000 males. In Delhi, Chadha et al. (1990) carried out a community-based study of 13,723 adults in the age group of 25 to 64. The diagnosis of coronary heart disease was based on clinical history supported by documentary evidence of treatment in a hospital or at home, or on ECG evidence in accordance with the Minnesota code. The total prevalence rate of coronary artery disease based on both criteria was 96.7/1000 adults. Three years later, a follow-up study of the original cohort (n = 4151), who were free from heart disease initially, revealed 245 new cases of coronary heart disease, giving an incidence of 19.7 per 1000 adults per annum (Chadha et al. 1993). In a more recent survey, Reddy (1993) used more

Table 4. Burden of late-onset disorders in India [in millions]

Coronary artery disease (>30 yr)	7.06
Hypertension (>20 yr)	12.92
Diabetes mellitus (>15 yr)	7.41
Mental illness (schizophrenia, manic depressive psychosis, etc.)	26.66
Epilepsy	17.70
Mental retardation	17.10
Total	88.85

[a]Adapted from Verma (1986a,b).

specific electrocardiographic criteria and detected a prevalence of 41/1000 adults of both sexes in the age range of 35 to 64. The addition of history-based diagnostic criteria increased this figure to 76/1000 adults. Krishnaswamy et al. (1991) showed that over a 19-year period in Vellore the number of admissions from coronary artery disease increased form 4% in 1960 to 33% in 1989. Tantrarongroj and Nelson (1990) also reported a six-fold increase in the prevalence of acute myocardial infarction during a 28- year period in Bangkok Adventist Hospital. A higher incidence of acute myocardial infarction was observed in Indians in Bangkok as compared with Thais and Chinese. In Indonesia a study in Jakarta (Boedhi-Darmejoet al. 1990) of 2,073 people aged 25 to 64 revealed electrocardiographic evidence of myocardial infarction in 27/1000 persons, while 134/1000 persons had hypercholesterolemia (>6.5 mmol/L).

The operation of genetic factors is highlighted by the fact that ischaemic heart disease continues to have a high frequency among Asian Indians migrating to the United Kingdom, Trinidad and South Africa (Seedat 1990; Hughes 1990). McKeigue et al. (1989, 1992) also suggested that coronary artery disease rates are high in overseas South Asian groups of different geographic regions. These rates were not explained on the basis of elevated serum cholesterol, smoking or hypertension. They suggested that these communities are characterized by low plasma HDL cholesterol, high plasma triglyceride levels and high levels of non-insulin dependent diabetes, reflecting an underlying state of insulin resistance. Berg et al. (1991) have summarized the data on the genetic risk factors for coronary artery disease, but there are hardly any data on the prevalence of these factors in India.

Reddy (1993) recently determined the risk factor profile for cardiovascular disease in an urban and rural population in and around Delhi. In the urban population, the major risk factors were history of hypertension, diabetes mellitus, and coronary artery disease, as well as the presence of hypertension and diabetes mellitus found upon examination. Even a family history of hypertension, diabetes mellitus, or coronary artery disease was significant, as was the presence of hypercholesterolemia. In the rural population the major risk factors were smoking, the presence of hypertension found upon examination, and hypercholesterolemia.

Table 5. Urban distribution by subjects of number by modifiable coronary risk factors [%][a]

	Males			Females		
	35–44 years n = 475	45–54 324	55–64 305	35–44 years n = 515	45–54 329	55–64 302
0	34.8	23.8	22.6	58.8	37.7	22.2
1	44.8	47.2	38.4	33.6	39.2	40.4
2	16.0	21.9	29.2	6.6	19.5	31.1
3	4.2	7.1	9.5	0.8	3.6	6.3
4	0.2	0.0	0.3	0.2	0.0	0.0

[a]Adapted from Reddy (1993).

Table 5 shows the distribution of subjects by the number of modifiable coronary risk factors. The presence of risk factors increased by age. One risk factor was identified in 38-44% of men, and 34-41% of women of different age groups. Two risk factors were present in 16-29% of men and 6.6-31% of women of different ages.

Hypertension

The data on the prevalence of hypertension as determined by community surveys in India are summarised in Table 6. In earlier surveys based on the criteria of systolic blood pressure exceeding 160 mm Hg and diastolic blood pressure exceeding 95 mm Hg, the prevalence among urban subjects 20-60 years of age was 5.9% in males and 6.99% in females (Gupta et al. 1978). More recent results on prevalence of hypertension in Delhi using the criteria of systolic/diastolic pressure of more than 140/90 mm Hg have varied from 11.6% in males and 13.6% in females (Chadha et al. 1990), to 17.4% in both sexes combined (Reddy 1993). Gopinath et al. (1994) resurveyed after 3 years the population screened earlier by Chadha et al. (1990) and documented the incidence of hypertension as 12.0 per 1000 for both males and females.

The different studies show that the prevalence increases with age, socioeconomic status, urbanisation and stressful living. The risk factors for

Table 6. Prevalence of hypertension in India

Place	Age [yr]	Criteria	%		Author (period of study)
			M	F	
Rohtak (U)[a]	20-60	WHO criteria 165/95 mm Hg	5.99	6.99	Gupta et al. (1978)
Haryana (R)		WHO criteria 165/95 mm Hg	3.55	3.59	Gupta et al. (1977)
Delhi	35-64	>140/90mm Hg	17.4	(C)	Reddy (1993-94)
Delhi	25-64	>140/90mm Hg	11.66	13.68	Chadda (1985-87)

[a]U, urban; R, rural; C, Combined sexes.

Table 7. Risk factors for hypertension in India [%][a]

	Cases n = 132	Controls n = 3,289
Smoking	7.6	12.4
Family history	12.1	14.5
Obesity	24.2	22.1
Diabetes mellitus	9.8*	1.0
Alcohol consumption	5.3*	1.2

[a]Adapted from Gopinath et al. (1994).

hypertension, as determined by a recent study by Gopinath et al. (1994) in Delhi, are summarised in Table 7. A comparison of cases and controls showed a significant increase only in alcohol consumption and the presence of diabetes mellitus.

Diabetes Mellitus

The frequency of diabetes mellitus is increasing in the developing countries (King et al. 1991). The prevalence of diabetes mellitus, as revealed by studies in different parts of India, is summarised in Table 8. The prevalence varies from 0.5% to 8.2% in different parts of the country. In a well-designed study in Madras (Ramachandran 1992), the age-adjusted prevalence of diabetes mellitus in an urban population was 8.2%, while impaired glucose tolerance test was detected in another 8.7%. In the rural population the prevalence of diabetes mellitus was much lower (2.4%), while that of impaired glucose tolerance was similar (7.8%).

The high frequency of diabetes mellitus among Indians persists even after migration to foreign countries (Table 9). The striking fact that emerges from this table is that Indians living in foreign countries have a higher frequency than other ethnic groups resident in those countries. This is true for Trinidad, Guyana and South Africa (Indians vs. Negroes), Singapore (Indians vs. Chinese and Malays), and the Fiji Islands (Indians vs Melanesians). It reflects the operation of genetic factors that predispose the Indians to a high frequency of diabetes mellitus. However, no data are available on the predisposing genetic factors among the Indians. Limited studies do show a positive correlation with the following factors: urban population, age, body mass index, waist-hip ratio, intake of refined food low in fibre, low utilisation of calories due to a lack of physical activity, presence of hyperinsulinemia and insulin resistance. A hypothesis for the development of glucose intolerance and NIDDM among Asians is set out in Fig. 4.

Table 8. Prevalence of diabetes mellitus in India [%][a]

Year	Author	Place	Urban	Rural
1971	Tripathy et al.	Cuttack	1.2	-
1972	Ahuja et al.	New Delhi	2.3	-
1979	Johnson et al.	Madurai	0.5	-
1979	Gupta et al.	Multicentre	3.3	1.3
1984	Murthy et al.	Tenali	4.7	-
1986	Patel	Bhadran	-	3.8
1988	Ramachandran et al.	Kudremukh	5.0	-
1989	Kodali et al.	Gangavathi	-	2.2
1989	Rao et al.	Eluru	-	1.6
1992	Ramachandran et al.	Madras	8.2	2.4

[a]Adapted from Ramachandran (1992).

Table 9. Prevalence of diabetes mellitus in migrant Asian Indians[a]

Country	Population studied	Males	Females	Totals
Trinidad	Indians	2.3	1.0	1.7
(1958)	Negroes	1.4	1.5	1.4
(1968)	Indians	2.5	2.3	4.5
	Negroes	1.0	2.1	2.5
	Mixed	1.2	1.5	4.4
Guyana	Indians			5.7
(1962)	Africans			0.6
South Africa	Indians			10.4
(1969)	Malays			6.6
	Africans			3.6
Singapore	Indians	8.1	3.1	6.1
(1975)	Chinese	1.7	1.4	1.6
	Malays	2.7	2.2	2.4
Fiji	Indians (U)[b]	12.9	11.0	
	Melanesians (U)	3.5	7.1	
	Indians (R)	12.1	11.3	
	Melanesians (R)	1.1	1.2	

[a]Adapted from Ramaiya et al. (1990).
[b]U, urban; R, rural.

Syndrome X Among Indians

Evidence is accumulating both from studies abroad (McKeigue et al. 1991) and in India (Misra 1994) that syndrome X, or insulin resistance syndrome, has a high frequency among Indians. This syndrome is characterized by:

1) insulin resistance, which is the central feature;
2) hyperinsulinemia;
3) hypertriglyceridemia;
4) low high density lipoproteins;
5) hypertension; and
6) obesity, mostly, intra abdominal with elevated waist to hip ratio.

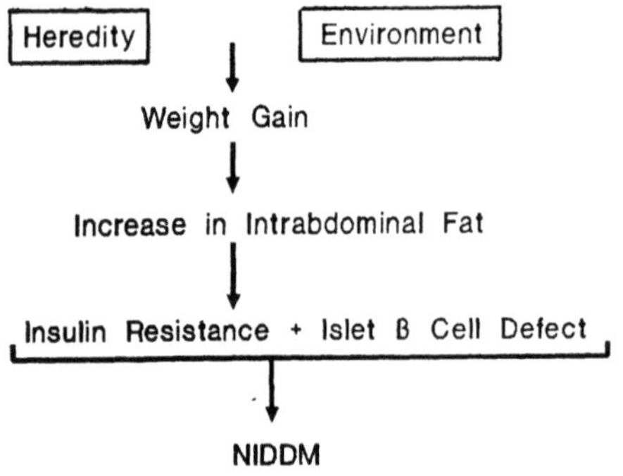

Fig. 4. Hypothesis for the development of glucose intolerance and non-insulin dependent diabetes mellitus (NIDDM)

Due to insulin resistance there is a failure to suppress release of nonesterified fatty acids from adipose tissue. This leads to high triglycerides and low high density lipoproteins. What is the genesis of syndrome X? Obviously hereditary and environmental factors interact, and under increased dietary intake weight gain ensues. This leads to accumulation of intrabdominal fat. The presence of insulin resistance and impairment of islet beta cell function results in non-insulin dependent diabetes mellitus and cardiovascular disease.

An interesting hypothesis has been advanced by Barker and colleagues (Barker 1993a,b). Subjects born in Sheffield and Preston between 1907 and 1925 and in Hertfordshire between 1911 and 1930 were followed into adult life, and weight at birth and at one year was correlated with the presence of complex disorders and syndrome X (Barker et al. 1989). Standardised mortality due to coronary artery disease was 111 in those men who had weighed 8.2 kg at one year, and 42 in those who had weighed 12.3 kg or greater at one year. The association was less strong with low birth weight. The frequency of diabetes mellitus was 27% in those with a birth weight of 5.5 lbs or less, and 6% in those, with a birth weight of 7.5 lbs or more. The frequency of syndrome X was 30% in those weighing 5.5 lbs or less at birth, as compared to 6% in those who weighed 9.5 lbs or more at birth (Barker et al. 1993).

These observations indicate that syndrome X, commonly prevalent in migrant and relatively affluent urban Indians, may be the result of low rates of fetal and infant growth. The effect was found in babies who were born small for their gestational age, rather than in those born prematurely. The effect was not confined to babies with intrauterine growth retardation but was also seen in babies of average or above average weight. Some of these babies were small in relation to the size of their placentas, or were short in relation to their heads. In others, average birth weight was followed by below average infant weight gain.

Barker and colleagues suggested that processes associated with low rates of fetal and infant growth programme cardiovascular disease and diabetes. They quote numerous animal experiments that show that poor nutrition and other influences that reduce growth during critical periods of early life may permanently affect the structure and physiology of a range of organs and tissues. The long-term consequences of early growth restraint depend on its timing, because different tissues mature at different stages of fetal size and infancy. Consistent with this hypothesis, babies with different patterns of reduced fetal growth have different abnormalities in adults. For example, blood pressure was related to birth weight, but not independently to weight at one year. In contrast, plasma total and low density liproprotein and cholesterol concentrations were related to the method and duration of infant feeding. The relation to early growth constraint was found to be strong, graded and specific. If true, these observations have wide-ranging implications in developing countries, where the prevalence of low birth weight is extremely high. It is possible that malnutrition during critical phases of growth (intrauterine life and infancy) could result in organ and tissue damage of a selective type, which manifests in later life as disease after exposure to an affluent life style. This is an additional, albiet critical, risk factor for adult diseases. Geneticists can perhaps take the correct perspective that predisposed individuals with a certain genetic constitution would have a higher risk of developing syndrome X when exposed to adverse environmental risk factors in early or later life. Obviously more will be heard about this in the near future.

Conclusions

Urbanisation is proceeding at a very rapid pace in developing countries. The process of urbanisation is changing the life style of the people. There is less physical activity accompanied by increased intake of refined foods and saturated fats and greater stress. These environmental factors, obviously acting with the genes in the predisposed subjects, are leading to a high frequency of coronary artery disease, hypertension and diabetes mellitus. Geneticists are expending a lot of effort in searching for the genes predisposing to these common disorders, while the preventive cardiologists are engaged in modifying environmental and life style factors to reduce the morbidity and mortality related to these disorders. With greater understanding of the genetic component in the etiology of these disorders, we will be better able to identify the environmental factors. More importantly there would be no need to recommend modification of environmental factors to every one, which has not worked so far, but learn to recommend the modification in a selective fashion on those who are genetically predisposed, an approach which is likely to be successful.

References

Barker DJP (1993a) Fetal origins of coronary heart disease. Br Med J 69: 195–196

Barker DJP (1993b) Intrauterine growth retardation and adult disease. Curr Obstet Gynaecol 3: 200–206

Barker DJP, Winter PD, Osmond C, Margetts B, Simmonds SJ (1989) Weight in infancy and death from ischaemic heart disease. Lancet 9: 577–580

Barker DJP, Hales CN, Fall CHD, Osmond C, Phipps K, Clark PMS (1993) Type 2 (non-insulin dependent) diabetes mellitus, hypertension, and hyperlipidaemia (syndrome X) relation to reduced foetal growth. Diabetologia 36: 62—67

Berg K, Bulyzhenkov V, Christen Y, Corvol P (eds) (1991) Genetic approaches to coronary heart disease and hypertension. Springer Verlag, Berlin, pp 98–143

Boedhi-Darmoja R, Setianto B, Sutedjo, Kusmana D, Anradi, Supari F, Salan R (1990) A study of baseline risk factors for coronary heart disease - results of population screening in a developing country. Rev Epidemical Sante Publique 38: 487–91

Bower C, Stanley FJ, Croft M, De Klerk NH, Davis RE, Nicol DJ (1993) Absorption of pteroylopoly-glutamates in mothers of infants with neural tube defects. Br J Nutr 69: 827–834

Czeizel AE, Dubas I (1992) Prevention of the first occurrence of neural tube defects by periconceptional vitamin supplementation. New Engl J Med 327: 1832–1835

Chadha SL, Radhakrishnan S, Ramachandran K, Kaul U, Gopinath N (1990) Prevalence, awareness and treatment status of hypertension in urban population in Delhi. Indian J Med Res 92: 424–430

Chadha SL, Radhakrishnan S, Ramachandran K, Kaul U, Gopinath N (1990) Epidemiological study of coronary heart disease in urban population of Delhi. Indian J Med Res 92: 424–430

Chadha SL, Ramachandran K, Shekhawat S, Tandon R, Gopinath N (1993) A 3-year follow up study of coronary artery disease in Delhi. Bull WHO 71: 67–72

Copeman JB, Cucca F, Hearne CM, Cornall RJ, Reed PW, Rohnigen KS et al. (1995) Linkage disequilibrium mapping of a type 1 diabetes susceptibility gene (IDDM7) to chromosome 2q31–q33. Nature Genet 9: 80–85

Copp AJ (1994) Birth defects: from molecules to mechanisms. J Roy Coll Phys Lond 28: 294–300

Gopinath N, Chadha SL, Shekhawat S, Tandon R (1994) A 3-year follow up of hypertension in Delhi 72: 715–720

Gupta SP, Siwach SB, Moda VK (1977) Epidemiology of hypertension, based on total community survey in the rural population of Haryana. Indian Heart J 29: 53–62

Gupta SP, Siwach SB, Moda VK (1978) Epidemiology of hypertension based on total community survey in the urban population of Haryana. Indian Heart J 30: 315–322

Hughes LO (1990) Insulin, Indian origin and ischaemic heart disease. Int J Cardiol 26: 1–4

Indian Council of Medical Research Human Genetics Task Force Report on "Genetic Counselling and Prenatal Diagnosis" (1992) Indian Council of Medical Research, New Delhi

King H, Rewers M, WHO Ad Hoc Diabetes Reporting Group (1991) Diabetes in adults is now a third world problem. Bull WHO 69: 643–648

Krishnaswamy S, Joseph G, Richand J (1991) Demands on tertiary care for cardiovascular disease in India – analysis of data for 1960–1989. Bull WHO 89: 325–30

McKeigue PM (1992) Coronary heart disease in Indians, Pakistanis and Bangladeshis: aetiology and possibilities for prevention. Br Heart J 67: 341–342

McKeigue PM, Milker GJ, Marmot MG (1989) Coronary heart disease in South Asians overseas – a review. J Clin Epidemiol 42: 597–609

McKeigue PM, Shah B, Marmot MG (1991) Relation of central obesity and insulin resistance with high diabetes prevalence and cardiovascular risk in South Asians. Lancet 332: 382–386

Mendis S (1992) Prevention of coronary heart disease – putting theory into practice. Ceylon Med J 37: 9–11

Ministry of Health and Family Welfare (1994) Annual Report 1993–94. Government of India, New Delhi

Misra A (1994) A tale of two syndromes X. Natl Med J India 7: 26–27

Nelson K, Holmes LB (1989) Malformations due to presumed spontaneous mutations in newborn infants. New Engl J Med 320: 19–23

Newmann PE, Frankel WN, Letts VA, Coffin JM, Copp AJ, Bernfield M (1994) Multifactorial inheritance of neural tube defects: localisation of the major gene and recognition of modifiers in ct mutant mice. Nature Genet 6: 357–362

Ramachandran A (1992) Genetic epidemiology of NIDDM among Asian Indians. Ann Med 24: 499–503

Ramaiya KL, Kodali VRR, Alberte KGMM (1990) Epidemiology of diabetes in Asians of the Indian subcontinent. Diabetes/Metab Rev 6: 125–146

Reddy KS (1993) Cardiovascular disease in India. World Health Statist Quart 46: 101–107

Romeo G, McKusick V (1994) Phenotypic diversity, allelic series, and modifier genes. Nature Genet 7: 451–452

Scriver C (1992) Genetic diseases – effects on human health. In: Kuliev A, Greendale K, Penchazadeh V, Paul NW (eds) Genetic services provision – an international perspective. March of Dimes Birth Defects Foundation, White Plains, New York 1–16

Seedat YK (1990) Hypertension and vascular disease in India and migrant Indian populations in the world. J Hum Hypertens 4: 421–24

Sellers MJ (1983) The cause of neural tube defects: some experiments and a hypothesis. J Med Genet 20: 164–168

Smithells RW, Shepard S, Schorah CJ (1980) Possible prevention of neural tube defects by periconceptional vitamin supplementation. Lancet 1: 339–340

Strachan T, Read AP (1994) PAX genes. Curr Opin Genet Devel 4: 427–438

Tantrarongroj K, Nelson ER (1990) Incidence of first myocardial infraction in the Bangkok Adventist Hospital (1958–1985). J Med Assoc Thailand 73: 29–34

Thomson C (1994) Identifying complex disease genes progress and paradigms. Nature Genet 8: 108–110

Verma IC (1986a) Genetic counselling and control of genetic disease in India. In: Verma IC (ed) Genetic research in India. Sagar Printers and Publishers, New Delhi, pp 21–37

Verma IC (1986b) Genetic disorders need more attention in developing countries. World Health Forum 7: 69–70

Verma IC, Elango R, Mehta L (1990) Monitoring reproductive and developmental effects of environmental factors in India – a review. Abstracts Indo-US Symposium. Effect of environmental and genetic factors on pregnancy outcome. All India Institute of Medical Sciences, New Delhi, pp 63–73

Wald N, Sneddon J, Densen J, Frost C, Stone R, MRC Vitamin Study Research Group (1991) Prevention of neural tube defects: Results of the Medical Research Council vitamin study. Lancet 338: 131–137

Wilson GN (1993) Relevance of the genetic of embryologic development. Growth, Genet Hormones 9: 1–5

Subject Index

Springer-Verlag and the Environment

We at Springer-Verlag firmly believe that an international science publisher has a special obligation to the environment, and our corporate policies consistently reflect this conviction.

We also expect our business partners – paper mills, printers, packaging manufacturers, etc. – to commit themselves to using environmentally friendly materials and production processes.

The paper in this book is made from low- or no-chlorine pulp and is acid free, in conformance with international standards for paper permanency.

MIX
Papier aus verantwortungsvollen Quellen
Paper from responsible sources
FSC® C105338

If you have any concerns about our products,
you can contact us on
ProductSafety@springernature.com

In case Publisher is established outside the EU,
the EU authorized representative is:
Springer Nature Customer Service Center GmbH
Europaplatz 3, 69115 Heidelberg, Germany

Printed by Libri Plureos GmbH
in Hamburg, Germany